Modular Form

A classical and computational introduction

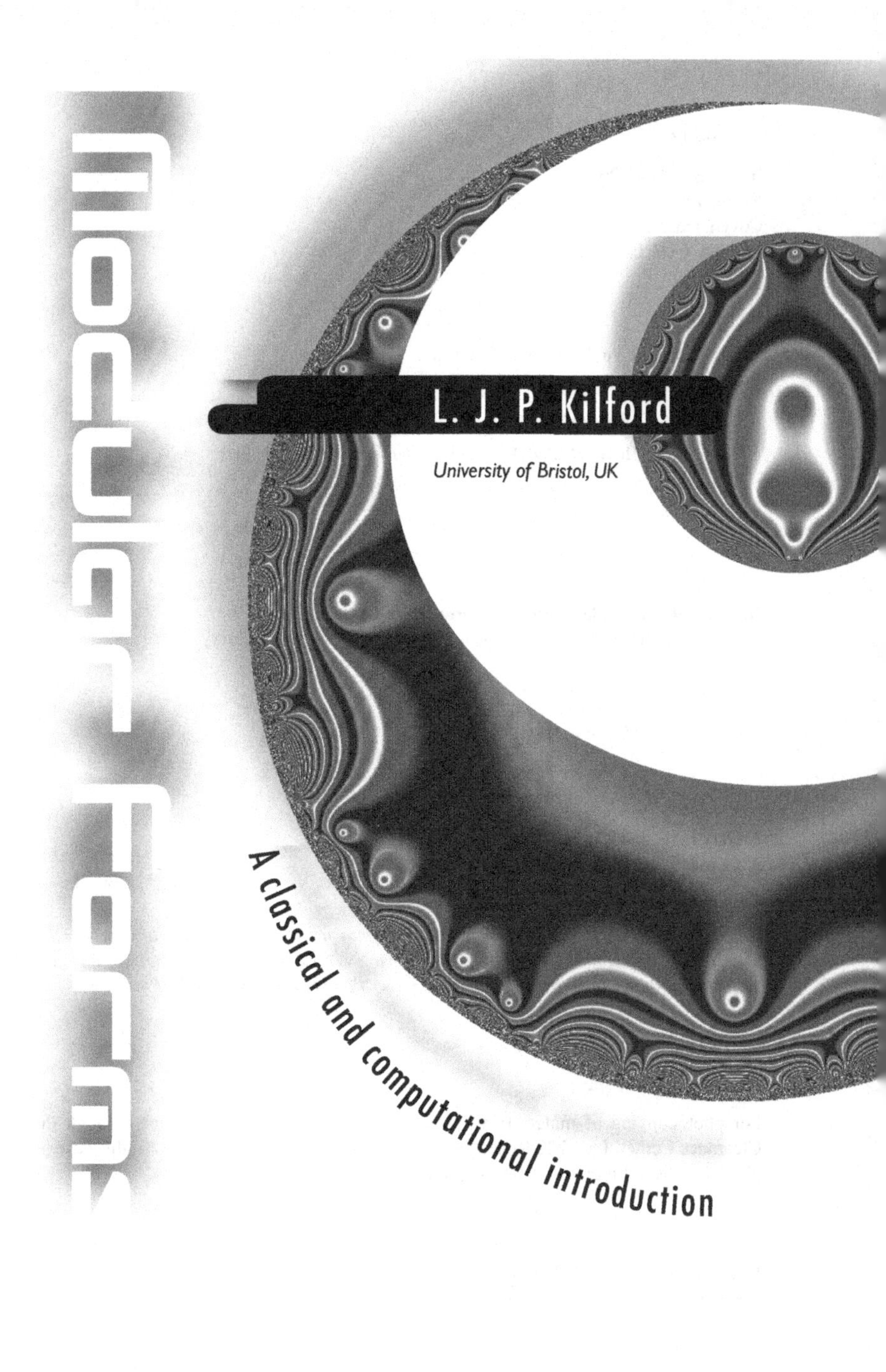

ICP
Imperial College Press

Published by

Imperial College Press
57 Shelton Street
Covent Garden
London WC2H 9HE

Distributed by

World Scientific Publishing Co. Pte. Ltd.
5 Toh Tuck Link, Singapore 596224
USA office: 27 Warren Street, Suite 401-402, Hackensack, NJ 07601
UK office: 57 Shelton Street, Covent Garden, London WC2H 9HE

British Library Cataloguing-in-Publication Data
A catalogue record for this book is available from the British Library.

MODULAR FORMS
A Classical and Computational Introduction

ISBN-13 978-1-84816-213-6
ISBN-10 1-84816-213-8

Printed in Singapore.

For my family, with thanks for everything

Acknowledgements

This book was written while I was the GCHQ Research Fellow at Merton College, in the University of Oxford and a Research Fellow at the University of Bristol. I would like to thank Merton College and the Heilbronn Institute for their hospitality and for providing pleasant environments to do research in.

I would like to thank Roger Heath-Brown, George Walker and Jahan Zahid of the University of Oxford for their participation in the course of 2006–2007, and I would especially like to thank Jahan Zahid and George Walker for reading earlier drafts of this book and suggesting many improvements; I would also like to thank Robin Chapman of the University of Exeter and Gabor Wiese of the Universität Duisburg-Essen for reading the manuscript and making helpful comments. Any remaining mistakes are due to the author.

I would like to thank Tomas Boothby for his assistance in creating the cover illustrations using SAGE; it is derived from pictures showing the absolute value of certain modular forms under the Cayley projection to the unit circle.

I would also like to thank Jeff Allotta of Northwestern University, Nathan Ryan of Bucknell University and William Stein of the University of Washington for helpful conversations.

Thanks are also due to my PhD supervisor, Kevin Buzzard of Imperial College London, for being an inspiration, both during my PhD and afterwards.

Taruith in Oxford and ICSF in London provided much-needed distraction while this book was being written; I offer my thanks to them.

I would also like to thank Vivienne for everything.

Contents

Introduction

This book is based on notes for lectures given at the Mathematical Institute at the University of Oxford over the three years 2004—2007; as a graduate course both in 2004—2005 and in 2006—2007, and as the undergraduate course *Introduction to Modular Forms* in 2005—2006.

This book

This book focuses on the computational aspects of the theory of modular forms much more than most other books do. It is designed to be an introduction to these computational aspects of the theory. Computational algebra packages like MAGMA and SAGE can (for instance) compute Fourier expansions of modular forms and modular functions to an extremely high precision; this book will hopefully help the student to use computers to deepen their understanding of modular forms and number theory.

It also gives a grounding in the theory of classical modular forms, starting with modular forms for $\mathrm{SL}_2(\mathbf{Z})$, and progressing to modular forms of higher level, modular functions, half-integral weight modular forms and mod p modular forms. The two aspects are intended to complement each other, each helping to motivate and inspire the other.

Possible courses

The courses that this book is based upon were all 16-lecture courses, given over one Oxford term. The undergraduate course was given to final-year undergraduate students and to students taking a one-year taught Master's course.

This book concentrates on the computational aspects of the theory of modular forms, but at the same time also gives a grounding in the classical and theoretical aspects of modular forms. We now list some possible courses which could be given based upon this book.

One possible way to give a course using this book would involve teaching students essentially only about modular forms for the full modular group, as [Serre (1973a)] does, in a broadly "traditional" way. This could still include interesting arithmetic, such as a study of the Ramanujan τ function and the j-invariant, but avoids the complication of higher levels. There are some projects in the computational chapter which could be used to stretch the more able students and also serve as inspiration for end-of-course projects, especially for beginning graduate students.

If the instructor has more time, then the theory of modular forms for congruence subgroups of $\mathrm{SL}_2(\mathbf{Z})$ could also be introduced. This would allow the introduction of many well-known modular forms, such as those associated to elliptic curves, many η-products, and theta functions. The work of Wiles and Taylor in proving the modularity conjecture, and hence Fermat's Last Theorem, is of great public interest, and the general concepts of the proof can be presented.

Another possible course would be one focused on introducing students to computational number theory, using modular forms as motivational examples. This might include teaching students about MAGMA, PARI and SAGE, as well as giving them an overview of the history of computation. End-of-course projects for this course could include more or less programming to suit the instructor and the level of the students. The graphically inclined could write programs similar to those of Verrill [Verrill (2001)] to create pictures of fundamental domains, or use the graphics creation facilities of PARI and SAGE to create other graphics.

A third possibility would be to teach a course dealing with theta functions, starting with their history (see, for example, [Glaisher (1907a,b,c)], for early computational results) and continuing to modern-day work. This course could balance theoretical work showing that theta functions really are modular forms with computations of explicit theta functions, and more advanced students could read books like [Miyake (2006)] which give more details on this subject.

The instructor could also teach a course emphasizing the applications of modular forms in number theory and elsewhere. It is important to realize that, despite their definition as holomorphic functions from $\mathcal{H}$ to $\mathbf{C}$, modular forms have a rich number-theoretic theory, and Chapter 5 gives a sample of some of the many and varied results that rely on the theory of modular forms.

An overview of this book[1]

We encourage the reader of this book to use their favourite computer algebra package (such as MAGMA or SAGE) to compute examples as they read, to help them develop their intuition. The best way to learn how to use such a package is to experiment with it to find out how it works; Chapter 7 gives an introduction and brief overview, but there is no substitute for hands-on experience.

The chapters cover the theory in the following way; Chapter 2 gives the definition of a modular form for $\mathrm{SL}_2(\mathbf{Z})$, and prove that modular forms exist (which is not obvious to the newcomer to the subject). We also define fundamental domains and modular forms for congruence subgroups.

In Chapter 3, we consider modular forms as complex-analytic objects, and use this aspect of their character to prove that spaces of modular forms of a given weight for a given congruence subgroup are finite-dimensional. Following on from this, in Chapter 4 we introduce the Hecke operators, which are linear operators acting on spaces of modular forms, and prove results about them. We note that these operators explain the recurrence relations that the coefficients of $\Delta(q)$ satisfy, for instance.

We then apply the results derived in these chapters in Chapter 5 to a variety of applications, such as Fermat's Last Theorem, computing digits of π, and computing the number of representations of integers by quadratic forms. We also introduce the concept of mod p modular forms in Chapter 6, give structure theorems for mod p modular forms, and talk about Serre's Conjecture, which has been proved very recently.

Finally, we consider the practical side of computation in Chapter 7; after giving a brief introduction to the history of computations in the world of number theory, which includes such highlights as the Lehmer bicycle-chain sieve, we introduce the computer algebra packages MAGMA, SAGE and PARI, and briefly touch on the theoretical side of computing in mathematics. The book ends with appendices containing examples of code for the algebra packages discussed in the text.

[1]Many books have a section of this nature. We note that [Lamport (1994)] is one of the very few that has a section called "How to Avoid Reading This Book".

Chapter 1

Historical overview

> The study of the history of mathematics will not make better mathematicians but gentler ones, it will enrich their minds, mellow their hearts, and bring out their finer qualities. — George Sarton.

We will introduce historical topics in the main text as appropriate; in this chapter we will give a broad outline and overview of the history of research in the area of modular forms, in order to give a grounding in the subject. In Chapter 7, we will give a historical account of computation in number theory generally, with an emphasis on modular forms-related topics.

1.1 18th Century — a prologue

In [Hellegouarch (2002)], Section 5.1, a brief summary of the history of modular forms is given; he begins with the work of Jakob Bernoulli[1]. His book *Ars Conjectandi*, published posthumously in 1713, contains theta functions such as

$$\sum_{n=0}^{\infty} m^{n^2} \text{ and } \sum_{n=0}^{\infty} m^{\frac{n(n+3)}{2}}.$$

These will turn out to be intimately connected to the theory of modular forms.

We note that although there are arithmetic functions which play a role in the theory of modular forms which were considered long before the time

[1] He is also known as Jakob I Bernoulli in the literature; there were many members of the Bernoulli family active in mathematics.

of Bernoulli, such as figurate numbers (which we will meet again in Section 5.10), they were not studied in the context of modular forms.

Theta functions appear also in the work of Leonhard Euler, in connection with identities for the partition function. Let $p(n)$ be the number of partitions of n; in other words, the number of ways that we can write an integer n as a sum of positive integers. For instance, $2+2$ is a partition of 4, and it can be shown by explicit enumeration that $p(4) = 5$. It has long been known that $p(n)$ goes up very quickly as n increases; in the early years of the 20th century, Percy MacMahon proved that $p(200) = 397299029388$ using a recurrence relation (see [Hardy and Ramanujan (1917)], Section 1.7). Euler proved, amongst other things, that

$$\sum_{n=0}^{\infty} p(n)q^n = \prod_{m=1}^{\infty} \frac{1}{1-q^m}.$$

We will see in Chapter 5 that the right-hand side of this is essentially the reciprocal of one important building block for modular functions, the Dedekind η function. The equating of power series with infinite products will prove to be a recurrent theme in the theory of modular forms.

1.2 19th century — the classical period

Carl Jacobi developed the theory of theta functions and used this to prove, amongst other results, the following theorem ("Jacobi's four-squares theorem"):

$$\left(\sum_{n=-\infty}^{\infty} q^{n^2}\right)^4 = \sum_{a,b,c,d\in\mathbf{Z}} q^{a^2+b^2+c^2+d^2} = 1 + 8\sum_{m=1}^{\infty}\sum_{\substack{d\mid m\\ 4\nmid d}} dq^m.$$

We will see later that this can be interpreted as an equality of modular forms of level $\Gamma_0(4)$ and weight 2; indeed, once the correct machinery has been set up, the proof will be very short.

In the later 19th century, Felix Klein used function-theoretic methods to investigate certain modular functions; for instance, the j-invariant, which we will define in Chapter 4, is also called "Klein's absolute invariant" or "Klein's modular function" in older texts, because Klein defined it, gave an explicit formula for it, and showed that it was invariant under the action of the modular group $\mathrm{SL}_2(\mathbf{Z})$. The j-invariant is used in the theory of elliptic curves, which we will briefly touch on later, as an invariant; it classifies elliptic curves over $\mathbf{C}$ up to isomorphism.

1.3 Early 20th century — arithmetic applications

In the early 20th century, Srinivasa Ramanujan studied the explicit Fourier expansions of certain well-known modular forms, such as the Δ function, given by

$$\Delta(q) := q \cdot \prod_{n=1}^{\infty} (1-q^n)^{24} = \sum_{n=1}^{\infty} \tau(n) \cdot q^n,$$

and made a famous conjecture about the rate of growth of $\tau(n)$, which is known as the *Ramanujan τ function.* We will discover more of the properties of the Δ function throughout this book. He also generalized Jacobi's notion of a theta function, as well as discovering a new family of similar objects, called *Ramanujan mock theta functions.*

At about the same time, G. H. Hardy and J. E. Littlewood used results on the growth of Fourier coefficients of modular forms to bound $r_m(n)$, the number of ways that one can write n as the sum of m squares. This extends the work of Jacobi mentioned above; for n even, one shows that

$$\sum_{n=0}^{\infty} r_m(n) \cdot q^n$$

is a modular form for $\Gamma_0(4)$ of weight $m/2$, and then uses results about the size of coefficients of modular forms to obtain bounds. We will see in Chapter 5 how this works, and also consider the question of what happens when n is *odd*; this leads us into the fascinating world of half-integral weight modular forms.

In the 1920s and 1930s, Erich Hecke worked on the *Hecke algebras* associated to spaces of modular forms, and considered linear operators acting on spaces of modular forms; we now call these the *Hecke operators*, and we will consider the theory of these operators in Chapter 4. These generalized in a very useful way the observations of Ramanujan that the following formula holds:

$$\tau(m \cdot n) = \tau(m) \cdot \tau(n), \text{ if } (m,n) = 1,$$

and, if p is prime,

$$\tau(p^{k+1}) = \tau(p^k) \cdot \tau(p) - p^{11} \cdot \tau(p^{k-1}), \text{ for } k \geq 1.$$

We will see using results from Chapters 3 and 4 that these identities follow because the space of modular forms spanned by Δ is 1-dimensional, so it is automatically an eigenform for the Hecke operators.

In Bryan Birch's memoir of Oliver Atkin [Birch (1998)], he says that "[a]t the beginning of the century, modular forms were a vital branch of mathematics, as they are now; but between about 1930 and 1950 the theory was completely (if wrongly) out of fashion"; in the Foreword to Serge Lang's book on modular forms [Lang (1976)] the hiatus is attributed to the effects of the Second World War and interest instead being directed to "algebraic methods" in the theory of numbers.

During this period, there were some mathematicians still working in the field; Birch and Lang both mention Robert Rankin as the only person in Britain working on modular forms; he continued the study of the arithmetic properties of the Fourier coefficients of modular forms that Ramanujan had begun. We will prove a special case of one of Rankin's results at the end of Chapter 2.

In another direction, Atle Selberg considered non-holomorphic analogues of modular forms in [Selberg (1956)], which are called *Maass forms*. One recent application of these is to the theory of partitions; [Bringmann and Ono (2007)] show that the number of partitions of a positive integer can be obtained by considering a certain Maass form.

1.4 Later 20th century — the link to elliptic curves

Ramanujan had conjectured that, when p is a prime,

$$|\tau(p)| \leq 2p^{11/2}.$$

This conjecture was generalized by Hans Petersson to consider all modular cusp forms, and was proved by Pierre Deligne in the 1970s as a consequence of his proof of the Weil conjectures on algebraic varieties over finite fields.

In the 1950s, Goro Shimura and Yutaka Taniyama circulated a conjecture[2] relating Fourier coefficients of certain modular forms of weight 2 and the number of points of certain geometric objects called elliptic curves over finite fields. This was later worked on by André Weil, whose name is also associated with the conjecture.

It was proved in complete generality in 1999 by Breuil, Conrad, Diamond and Taylor [Breuil *et al.* (2001)], building on the famous work of Andrew Wiles and Richard Taylor [Wiles (1995)], [Taylor and Wiles (1995)],

[2]This has been called the Shimura-Taniyama conjecture, the Taniyama-Shimura conjecture, the Taniyama-Weil conjecture, the Shimura-Taniyama-Weil conjecture or the Modularity conjecture.

who proved it in the important special case where the level of the modular form (equivalently, the conductor of the elliptic curve) was squarefree. This sufficed to prove Fermat's Last Theorem, that there are no nontrivial solutions over $\mathbf{Z}$ to $x^n + y^n = z^n$ when $n \geq 3$. For a one-volume overview of the proof of Fermat's Last Theorem, we refer the reader to [Cornell *et al.* (1997)]. We will briefly consider this in Chapter 5.

It should also be noted that one can use generalizations of this work to Hilbert modular forms (which we will not define here) to prove a version of Fermat's Last Theorem over $\mathbf{Q}(\sqrt{2})$; see [Jarvis and Meekin (2004)] for the details. This requires n to be at least 4, because there actually exist explicit nontrivial solutions over $\mathbf{Q}(\sqrt{2})$ when $n = 3$; for instance one can check explicitly that $(18 + 17\sqrt{2})^3 + (18 - 17\sqrt{2})^3 = 42^3$.

Another result that has very recently been proved is Serre's Conjecture, which says that every Galois representation which satisfies certain conditions can be associated to a modular form of a given weight and level. A special case of this result is used in the proof of Fermat's Last Theorem. We will discuss Serre's Conjecture in Chapter 6.

1.5 The 21st century — the Langlands Program

The inspiration for the Langlands Program can be taken to be the Artin reciprocity laws, which generalize quadratic reciprocity; these give relations between one-dimensional representations of Galois groups of algebraic number fields and generalizations of the Riemann ζ function called Dirichlet L-series. The slogan is that this is a relationship between "arithmetic" and "geometric" objects.

This can be generalized again to give a structure of relations that should hold between automorphic representations, on the one hand, and (modular) Galois representations on the other hand. This is a very active field of work, and much still remains to be done, but much of the theory is known for the special case of modular forms for GL_2, which are the modular forms that we will be dealing with in this book.

Some introductory references for the Langlands Program are [Bump *et al.* (2003)], [Knapp (1997)] and [Murty (1993)].

Chapter 2

Introduction to modular forms

> There are five fundamental operations of arithmetic: addition, subtraction, multiplication, division, and modular forms. – Martin Eichler (attributed).

In this chapter, we will introduce modular forms and some of their basic theory. This will comprise an introduction to modular forms for $\mathrm{SL}_2(\mathbf{Z})$ and for congruence subgroups, give examples to illustrate the theory, compute some Fourier expansions, and define fundamental domains. It will gives hints of the rich complex and arithmetic theory of modular forms which will be covered in later chapters.

There are several standard references for the basic theory of modular forms; Chapter III of [Koblitz (1993)], Chapter 1 of [Diamond and Shurman (2005)], Chapter 4 of Milne's online notes [Milne (1997b)], Chapter I of [Lang (1976)] and Chapter VII of [Serre (1973a)] are all good references, although Serre only deals with modular forms for $\mathrm{SL}_2(\mathbf{Z})$.——————————————————————————- — SAGE Version 3.0.1, Release Date: 2008-05-05 — — Type notebook() for the GUI, and license() for information. — ——————————————————————————-

The notation that these books use is not unified, as there are many choices to make and different authors have made different choices; we note in the text a few of the more unfortunate collisions of notation.

2.1 Modular forms for $\mathrm{SL}_2(\mathbf{Z})$

> The modular group ... is like an octopus, with tentacles reaching out into many branches of pure mathematics ... – Gareth Jones.

We define the *full modular group* $\mathrm{SL}_2(\mathbf{Z})$ (also known simply as *the modular group*) to be the group of 2×2 matrices with integer entries and determinant 1:

$$\mathrm{SL}_2(\mathbf{Z}) := \left\{ \begin{pmatrix} a & b \\ c & d \end{pmatrix} : a, b, c, d \in \mathbf{Z},\ ad - bc = 1 \right\}.$$

(The "SL" here stands for the Special Linear Group, as opposed to $\mathrm{GL}_2(\mathbf{Q})$, which is the General Linear Group). This group is occasionally called the *unimodular group* in the literature.

Some references use the notation Γ for $\mathrm{SL}_2(\mathbf{Z})$; it is quite common to write an element of $\mathrm{SL}_2(\mathbf{Z})$ as γ.

It is well-known that $\mathrm{SL}_2(\mathbf{Z})$ is finitely generated; a standard presentation (see Exercise 1) is $\langle S, T | S^2, (ST)^3 \rangle$, where $T := \left(\begin{smallmatrix} 1 & 1 \\ 0 & 1 \end{smallmatrix}\right)$ (T here stands for "translation") and $S := \left(\begin{smallmatrix} 0 & 1 \\ -1 & 0 \end{smallmatrix}\right)$. This group is called $\mathrm{PSL}_2(\mathbf{Z})$, the Projective Special Linear Group, which is sometimes called the *modular group* in the literature, as in the quote at the beginning of this section.

We can view elements of $\mathrm{SL}_2(\mathbf{Z})$ as acting in the following way on elements of the Riemann sphere $\widehat{\mathbf{C}} := \mathbf{C} \cup \{\infty\}$:

$$\begin{bmatrix} a & b \\ c & d \end{bmatrix} (z) = \frac{az + b}{cz + d}, \quad \text{for } z \in \mathbf{C}.$$

(We extend this to all of the Riemann sphere by defining $f(-d/c) = \infty$ and $f(\infty) = a/c$). These transformations are known as *Möbius transformations.*

We define $\mathcal{H}$ to be the *Poincaré upper half plane*; it is all of the complex numbers z which have $\Im(z) > 0$ (where $\Im$ denotes the imaginary part of z). We will see in Exercise 1b that the action of a matrix $M \in \mathrm{SL}_2(\mathbf{Z})$ on $\mathcal{H}$ is trivial if and only if $M = \pm \left(\begin{smallmatrix} 1 & 0 \\ 0 & 1 \end{smallmatrix}\right)$.

If $f : \mathcal{H} \to \mathbf{C}$ is a meromorphic function which satisfies the transformation formula

$$f\left(\frac{az + b}{cz + d}\right) = (cz + d)^k \cdot f(z) \text{ for } \begin{pmatrix} a & b \\ c & d \end{pmatrix} \in \mathrm{SL}_2(\mathbf{Z}) \text{ and } z \in \mathcal{H} \tag{2.1}$$

then we say that f is *weakly modular of weight k for* $\mathrm{SL}_2(\mathbf{Z})$. In particular, we have that

$$f(z + 1) = f(z), \tag{2.2}$$

$$f(-1/z) = (-z)^k f(z), \tag{2.3}$$

for all $z \in \mathcal{H}$, by substituting T and S into (2.1). Conversely, we see that if f is a meromorphic function from $\mathcal{H}$ into $\mathbf{C}$ which satisfies (2.2) and (2.3), then it is weakly modular for $\mathrm{SL}_2(\mathbf{Z})$, because T and S generate $\mathrm{SL}_2(\mathbf{Z})/\{\pm 1\}$; see Exercise 19 for the proof of this.

From (2.2), we see that f has a Fourier expansion of the following form in a suitable neighbourhood of the origin:

$$f(q) = \sum_{n \in \mathbf{Z}} a_n q^n, \text{ with } q := e^{2\pi i z}; \tag{2.4}$$

moreover, because f is meromorphic, only finitely many of the a_n with $n < 0$ are nonzero. We call $f(q)$ the *Fourier expansion of* f *at* ∞, or simply *the Fourier expansion of* f; it is often referred to as the q-expansion of f.

If $a_n = 0$ for $n < 0$, then we say that f *is holomorphic at* ∞. If $a_n = 0$ for $n \leq 0$, then we say that f *vanishes at* ∞.

Definition 2.1 (Modular forms for $\mathrm{SL}_2(\mathbf{Z})$). *Let k be an integer. We say that a meromorphic function $f : \mathcal{H} \to \mathbf{C}$ is a* modular form of weight k (for $\mathrm{SL}_2(\mathbf{Z})$) *if*

(1) f is weakly modular of weight k for $\mathrm{SL}_2(\mathbf{Z})$,
(2) f is holomorphic on $\mathcal{H}$, and
(3) f is holomorphic at ∞.

We also say that f is a modular form for the full modular group, *or that f* is a modular form for $\mathrm{SL}_2(\mathbf{Z})$. *Some authors say that f* has level 1.

The set of modular forms of weight k for the full modular group $\mathrm{SL}_2(\mathbf{Z})$ *is written as* $M_k(\mathrm{SL}_2(\mathbf{Z}))$.

We define the value of $f(z)$ at ∞ *to be the limit of $f(z)$ as $z \to i\infty$, and we write it as $f(\infty)$. If f vanishes at ∞ (so $f(\infty) = 0$) then we say that f is a* cusp form of weight k (for $\mathrm{SL}_2(\mathbf{Z})$). *The set of cusp forms is written* $S_k(\mathrm{SL}_2(\mathbf{Z}))$ *(some authors write it as* $M_k^0(\mathrm{SL}_2(\mathbf{Z}))$*).*

We now make some remarks about the terminology and concepts we have just introduced.

Remark 2.2. The set of cusp forms is called $S_k(\mathrm{SL}_2(\mathbf{Z}))$ because the German for "cusp form" is "Spitzenform". In French a cusp form is a "forme parabolique".

Remark 2.3. The modern terminology for weights is given above. In [Serre (1973a)], the terminology "modular form of weight $2k$" is used. In older books, modular forms of weight k are sometimes said to have weight $-k$, degree $-k$ or dimension $-k$; [Knopp (1970)] says that dimension was the older of the latter two terms. Also, sometimes in the past the notations $x :=$

$e^{2\pi iz}$ or $q := e^{\pi iz}$ have been used, but nowadays the notation q for $e^{2\pi iz}$ is almost universal (in [Armitage and Eberlein (2006)] the terminology q is used *both* for $e^{2\pi iz}$ and for $e^{\pi iz}$ at different points).

Remark 2.4. It is said that G. H. Hardy banned his students from using the term "modular form". Whatever the truth of this, it is noticeable that in papers such as [Hardy (1920)] and [Hardy and Ramanujan (1918)] he uses the terminology *elliptic modular function*, and does not call the Δ function a modular form.

Remark 2.5. The terms *elliptic modular form* and *integral modular form* are sometimes used to distinguish these modular forms from other types of modular form (for instance, mod p modular forms, which we will introduce in Chapter 6). These are different types of object; they are defined over different rings. In contrast, some authors do not insist that their modular forms be holomorphic on $\mathcal{H}$ or at ∞, and use the term *entire modular forms* for what we have defined as modular forms; these are different sorts of the same type of object, defined over the same ring and differing only in the conditions that they satisfy.

One highbrow way of defining the space of cusp forms of weight k is to say that it is the object $S_k(\mathrm{SL}_2(\mathbf{Z}))$ which makes the following sequence exact:

$$0 \longrightarrow S_k(\mathrm{SL}_2(\mathbf{Z})) \longrightarrow M_k(\mathrm{SL}_2(\mathbf{Z})) \xrightarrow{f \mapsto f(\infty)} \mathbf{C} \longrightarrow 0.$$

This encapsulates the definition given above.

We see that the zero function is always a modular form; the natural question is to ask are there any others? (For instance, in [Klingen (1990)], we find the quote "The existence of modular forms which are different from zero is by no means trivial".) Before we answer the question (in the affirmative), we will note a few simple consequences of the definition.

If we let $k = 0$, then we see that the transformation formula becomes

$$f\left(\frac{az+b}{cz+d}\right) = f(z),$$

and we see that any constant function f satisfies this, so $M_0(\mathrm{SL}_2(\mathbf{Z})) \supseteq \mathbf{C}$. In fact, we will see later (Proposition 3.3) that this is an equality; that every modular form of weight 0 for $\mathrm{SL}_2(\mathbf{Z})$ is a constant function. Conversely, from the definition we can show that a constant function is not a modular form of any nonzero weight.

We also see that if we substitute the matrix $\begin{pmatrix} -1 & 0 \\ 0 & -1 \end{pmatrix}$ into the transformation formula in (2.1), then we find that

$$f(z) = (-1)^k f(z), \text{ for all } z \in \mathcal{H}. \tag{2.5}$$

We see that if k is odd, then $f(z) = -f(z)$ for every $z \in \mathcal{H}$, and therefore that f is identically zero.

2.2 Eisenstein series for the full modular group

We now prove that the definition that we have given above is not vacuous, and that nonconstant modular forms do exist.

Proposition 2.6 (Eisenstein series). *Let k be an even integer which is at least 4 and let $z \in \mathcal{H}$. The function*

$$G_k(z) := \sum_{\substack{m,n \in \mathbf{Z} \\ (m,n) \neq (0,0)}} \frac{1}{(mz+n)^k} \tag{2.6}$$

is a nonzero modular form of weight k for $\mathrm{SL}_2(\mathbf{Z})$.

Proof. Proofs that a given function is a modular form often boil down to checking the three conditions given in Definition 2.1; showing that it is weakly modular, holomorphic on $\mathcal{H}$, and holomorphic at ∞. It can be useful to consider each of these individually.

First, we will show that $G_k(z)$ is holomorphic on $\mathcal{H}$ and at ∞. We will use Lemma 1 from Chapter VII of [Serre (1973a)]:

Lemma 2.7. *Let Ω be a lattice in $\mathbf{C}$. The series*

$$L := \sum_{0 \neq \rho \in \Omega} \frac{1}{|\rho|^t}$$

is absolutely convergent for $t > 2$.

Proof. [Proof of Lemma 2.7] We note that the number of elements of Ω such that $|\rho|$ is between the consecutive integers n and $n+1$ is $O(n)$, so the convergence of L is reduced to the convergence of $\sum_{r=0}^{\infty} r^{-(t-1)}$, where r is a positive integer, and we know that this converges for $t - 1 > 1$. □

Now let us suppose that z satisfies $|z| \geq 1$ and $|\Re(z)| \leq 1/2$; we will see in Section 2.5 that this region F is what is known as a *fundamental domain*, and that every point in $\mathcal{H}$ is equivalent via the action of $\mathrm{SL}_2(\mathbf{Z})$ to a point of F. We can then see that the following inequality holds:

$$\begin{aligned}|mz+n|^2 &= m^2 z\bar{z} + 2mn\Re(z) + n^2 \\ &\geq m^2 - mn + n^2 = |m\omega - n|^2,\end{aligned}$$

where $\omega = -1/2 + \sqrt{3}/2$. From the lemma, we see that this means that the series $\sum |m\omega - n|^-k$ (where we do not take $(m,n) = (0,0)$) is convergent, which means that the sum $G_k(z)$ converges normally in F, and also that $G(\gamma^{-1}z)$ converges normally in γF, for $\gamma \in \mathrm{SL}_2(\mathbf{Z})$. We note that the γF cover $\mathcal{H}$.

From the lemma and the discussion following it, we see that the sum in (2.6) is absolutely convergent and uniformly convergent on compact subsets of $\mathcal{H}$, so $G_k(z)$ is a holomorphic function on $\mathcal{H}$. In Proposition 2.6 we are using the assumption that $k \geq 4$ to ensure that the sum converges absolutely.

To show that $G_k(z)$ is finite at ∞, we will show that $G_k(z)$ approaches an explicit finite limit as $z \to i\infty$. The terms of $G_k(z)$ are of the form $1/(mz + n)^k$; those which have $m \neq 0$ will contribute 0 to the sum, while those which have $m = 0$ will each contribute $1/n^k$. Therefore we have

$$\lim_{z \to i\infty} G_k(z) = \sum_{0 \neq n \in \mathbf{Z}} \frac{1}{n^k} = 2\zeta(k),$$

which is finite (and nonzero).

To show that $G_k(z)$ is weakly modular for $\mathrm{SL}_2(\mathbf{Z})$, it will suffice to show that it transforms correctly under the matrices S and T; it can be seen that $G_k(z) = G_k(z+1)$ by substituting $z+1$ for z; we have already shown that $G_k(z)$ is uniformly and absolutely convergent so we can rearrange the terms as necessary. We now show that $G_k(z)$ transforms correctly under S by rearranging (2.3):

$$\begin{aligned} z^{-k} \cdot G_k(-1/z) &= \sum_{\substack{m,n \in \mathbf{Z} \\ (m,n) \neq (0,0)}} \frac{z^{-k}}{(-m/z + n)^k} \\ &= \sum_{\substack{m,n \in \mathbf{Z} \\ (m,n) \neq (0,0)}} \frac{1}{(-m + nz)^k} = G_k(z), \end{aligned}$$

as required (again, we are using the fact that $G_k(z)$ is uniformly and absolutely convergent on $\mathcal{H}$), and so therefore $G_k(z)$ is a modular form of weight k, which is what we wanted to prove. □

We see from the proof that $G_k(z)$ does *not* vanish at ∞, so we have an example of a nonzero form of nonzero weight which is not a cusp form. We will now construct our first example of a cusp form.

By Exercise 2, we see that the set $M_k(\mathrm{SL}_2(\mathbf{Z}))$ of modular forms of a given weight forms a complex vector space, and that the product of modular forms is again a modular form. Therefore we see that $G_k(z)^l$ is a modular form of weight kl, and $G_{k_1}(z) \cdot G_{k_2}(z)$ is a modular form of weight $k_1 + k_2$.

The explicit values of $\zeta(k)$, where k is a positive even integer, are well known; for instance $\zeta(4) = \pi^4/90$ and $\zeta(6) = \pi^6/945$. Therefore, we have that

$$\left(G_4^3\right)(\infty) = 8\zeta(4)^3 = \frac{\pi^{12}}{3^6 \cdot 5^3} \text{ and } \left(G_6^2\right)(\infty) = 4\zeta(6)^2 = \frac{4\pi^{12}}{3^6 \cdot 5^2 \cdot 7^2},$$

and therefore we see that if we define $D(z) := \frac{4}{49} \cdot G_4^3 - \frac{1}{5} \cdot G_6^2$, then we have that $D(\infty) = 0$. We will see in the next section that this function D is not identically zero, and so therefore is a nontrivial example of a cusp form.

2.3 Computing Fourier expansions of Eisenstein series

We will see later that the Fourier expansions of modular forms are very arithmetically interesting; for instance, many naturally occurring modular forms have Fourier coefficients which are multiplicative or satisfy recurrence relations. In this section we will begin our study of these Fourier coefficients by showing how to compute them.

From equation (2.4), we see that a modular form f has a Fourier expansion in some suitable neighbourhood of the origin. This is often called the *q-expansion of f (at ∞)* in the literature. We have also computed the coefficient a_0 of the Fourier expansion of the form $G_k(z)$. We will now exhibit the complete Fourier expansion of $G_k(z)$.

Proposition 2.8. *Let $k \geq 4$ be an even integer, and let $z \in \mathcal{H}$. The modular form $G_k(z)$ has Fourier expansion*

$$G_k(z) = 2\zeta(k) + 2\frac{(2\pi i)^k}{(k-1)!} \sum_{n=1}^{\infty} \sigma_{k-1}(n) q^n,$$

where we define $\sigma_{k-1}(n)$ to be the function

$$\sigma_{k-1}(n) := \sum_{0<m|n} m^{k-1}.$$

Proof. We will show this by some careful analysis of Fourier series for trigonometric functions; [Diamond and Shurman (2005)] say that "the reader who is unhappy with this unmotivated incanting of unfamiliar expressions for a trigonometric function should be reassured that it is a standard rite of passage into modular forms".

There is a formula for the cotangent function:

$$\pi\cot(\pi z) = \frac{1}{z} + \sum_{m=1}^{\infty}\left(\frac{1}{z+m} + \frac{1}{z-m}\right),$$

and we also have the identity

$$\pi\cot(\pi z) = \pi\frac{\cos(\pi z)}{\sin(\pi z)} = i\pi - \frac{2i\pi}{1-q} = i\pi - 2i\pi\sum_{n=0}^{\infty} q^n,$$

where $q := e^{2\pi i z}$. By equating these identities, we see that

$$\frac{1}{z} + \sum_{m=1}^{\infty}\left(\frac{1}{z+m} + \frac{1}{z-m}\right) = i\pi - 2i\pi\sum_{n=0}^{\infty} q^n. \tag{2.7}$$

We differentiate both sides of (2.7) $k-1$ times with respect to z to obtain the formula

$$\sum_{m\in\mathbf{Z}}\frac{1}{(m+z)^k} = \frac{(-2\pi i)^k}{(k-1)!}\sum_{n=1}^{\infty} n^{k-1}q^n, \tag{2.8}$$

which is valid for $k \geq 4$. We note that the left hand side of this looks very like a component of G_k, whereas the right hand side looks much like a component of the Fourier expansion given in the theorem. This does motivate our approach in retrospect; if one integrates both the definition of $G_k(z)$ in terms of the double sum and the Fourier expansion involving σ_{k-1}, then one arrives at something very similar to the two formulae for the cotangent function.

We will now use (2.8) to write $G_k(z)$ as a Fourier expansion. Because $k \geq 4$, we have absolute convergence of our series, so the following rearrangements are valid:

$$\begin{aligned}
G_k(z) &= \sum_{\substack{m,n\in\mathbf{Z}\\(m,n)\neq(0,0)}} \frac{1}{(mz+n)^k} && (2.9)\\
&= \sum_{0\neq n\in\mathbf{Z}}\frac{1}{n^k} + 2\sum_{m=1}^{\infty}\sum_{n=-\infty}^{\infty}\frac{1}{(mz+n)^k}, \quad \text{(rearranging)} && (2.10)\\
&= 2\zeta(k) + 2\frac{(2\pi i)^k}{(k-1)!}\sum_{m=1}^{\infty}\sum_{n=1}^{\infty} n^{k-1}q^{mn} \quad \text{(from (2.8))} && (2.11)\\
&= 2\zeta(k) + 2\frac{(2\pi i)^k}{(k-1)!}\sum_{n=1}^{\infty}\sigma_{k-1}(n)q^n \quad \text{(rearranging).} && (2.12)
\end{aligned}$$

$\square$

A standard notation for Eisenstein series is to write

$$E_k(z) := \frac{G_k(z)}{2\zeta(k)};$$

this is called the *normalized Eisenstein series of weight k (of level 1)*. For these modular forms, the following series identity holds:

$$E_k(q) = 1 - \frac{2k}{B_k}\sum_{n=1}^{\infty}\sigma_{k-1}(n)q^n,$$

where the B_k are the *Bernoulli numbers*, which are defined by

$$\frac{t}{e^t - 1} = \sum_{m=0}^{\infty} B_m \cdot \frac{t^m}{m!}.$$

The first few Bernoulli numbers are $B_1 = -1/2$, $B_2 = 1/6$, $B_3 = 0$ and $B_4 = -1/30$. It can be proved that $B_{2r+1} = 0$ for $r \geq 1$. We will investigate some of the arithmetic properties of the Bernoulli numbers in

One reason why the Bernoulli numbers are of independent arithmetic interest is because if a prime p does not divide the numerator of any of the Bernoulli numbers $\{B_2, B_4, \ldots, B_{p-3}\}$ then it is said to be a *regular prime* and Fermat's Last Theorem is known to be true for such exponents p (this is a result of Ernst Kummer). If a prime is not regular, it is said to be an *irregular prime*. It is not known if there are infinitely many regular primes, but it is not difficult to show that there are infinitely many which are *irregular* (this was originally proved by Jensen [Jensen (1915)] in a short paper in Danish; he actually showed that there are infinitely many irregular primes which are congruent to 3 modulo 4). Experimental evidence and heuristics suggest that about 61% of primes are regular; the irregular primes below 100 are 37, 59 and 67.

A more straightforward way of defining the E_k for $k \geq 4$ is by

$$E_k(z) = \frac{1}{2}\sum_{\substack{m,n\in\mathbf{Z}\\(m,n)=1}} \frac{1}{(mz+n)^k};$$

this can be shown by grouping together indices m and n in the sum defining $G_k(z)$ with a given greatest common divisor, again using the result that the sum is absolutely and uniformly convergent to allow us to rearrange the sum.

One motivation for considering normalized Eisenstein series in a computational context is that we can represent integers and rational numbers exactly, but we can only approximate the transcendental number π to a finite precision.

The Fourier expansions at ∞ of the first few nonzero normalized Eisenstein series are given by

$$E_4(q) = 1 + 240 \sum_{n=1}^{\infty} \sigma_3(n) q^n$$

$$E_6(q) = 1 - 504 \sum_{n=1}^{\infty} \sigma_5(n) q^n$$

$$E_8(q) = 1 + 480 \sum_{n=1}^{\infty} \sigma_7(n) q^n$$

$$E_{10}(q) = 1 - 264 \sum_{n=1}^{\infty} \sigma_9(n) q^n$$

$$E_{12}(q) = 1 + \frac{65520}{691} \sum_{n=1}^{\infty} \sigma_{11}(n) q^n$$

$$E_{14}(q) = 1 - 24 \sum_{n=1}^{\infty} \sigma_{13}(n) q^n.$$

By considering the constant term of E_{12}, we can prove that 691 is an irregular prime in the sense we defined above.

We note here that some authors normalize their Eisenstein series so that the Fourier coefficient $a_1 = 1$, to match the normalization used for cusp forms. This has the disadvantage that E_4 and E_6 no longer have integral Fourier expansions.

We can now verify that our cusp form $D = \frac{4}{49} \cdot G_4^3 - \frac{1}{5} \cdot G_6^2$ is not identically zero; the first term of its Fourier expansion has coefficient $256\pi^{12}/165375$ which is nonzero. Therefore there exists a nonzero cusp form.

We can compute this form much more cleanly by considering the normalized Eisenstein series $E_4(z)$ and $E_6(z)$; we define the Δ *function* and the *Ramanujan* τ *function* in the following way:

$$\Delta(z) := \frac{E_4(z)^3 - E_6(z)^2}{1728} = \sum_{n=1}^{\infty} \tau(n) q^n.$$

The Fourier coefficients of $\Delta(z)$ are all integers, and they are also multiplicative; that is, $\tau(m \cdot n) = \tau(m) \cdot \tau(n)$ if $(m, n) = 1$. They also satisfy recurrence relations; if p is a prime, then

$$\tau(p^n) = \tau(p) \cdot \tau(p^{n-1}) - p^{11} \cdot \tau(p^{n-2}), \text{ for } n \geq 2.$$

These facts are consequences of the theory of Hecke operators, which we will discuss in Section 4.1. We will see that Δ is a well-studied modular form, with interesting number-theoretic properties.

We assumed that $k \geq 4$ in the proof of Proposition 2.8. It is natural to ask what happens when $k = 2$. It turns out that the series that one would use to define $G_2(z)$ does not converge absolutely (basically because the harmonic series diverges), so the rearrangement we performed is no longer valid. We summarize the results that do hold:

Proposition 2.9. *There is a holomorphic function E_2 which has Fourier expansion at ∞*

$$E_2(q) = 1 - 24\sum_{n=1}^{\infty} \sigma_1(n)q^n,$$

which satisfies the following transformation formula:

$$z^{-2}E_2(-1/z) = E_2(z) + \frac{12}{2\pi i z}.$$

Proof. [Koblitz (1993)], Proposition III.2.7 and [Diamond and Shurman (2005)], Section 1.2 both carry out the necessary calculations. □

We will use the properties of E_2, which is not weakly modular, later to prove that certain functions *are* weakly modular. The function E_2 also plays an important role in the theory of mod p modular forms, which will be studied in Chapter 6, and we will briefly consider E_2 in the context of quasi-modular forms in Section 2.8.1.

2.4 Congruence subgroups

In this section, we will introduce some interesting subgroups of $\mathrm{SL}_2(\mathbf{Z})$, which will be useful for our study of modular forms.

Let N be a positive integer. We define the following subgroups of $\mathrm{SL}_2(\mathbf{Z})$:

$$\Gamma_0(N) := \left\{ \begin{pmatrix} a & b \\ c & d \end{pmatrix} \in \mathrm{SL}_2(\mathbf{Z}) : c \equiv 0 \mod N \right\} \tag{2.13}$$

$$\Gamma_1(N) := \left\{ \begin{pmatrix} a & b \\ c & d \end{pmatrix} \in \Gamma_0(N) : a \equiv d \equiv 1 \mod N \right\} \tag{2.14}$$

$$\Gamma(N) := \left\{ \begin{pmatrix} a & b \\ c & d \end{pmatrix} \in \Gamma_1(N) : b \equiv 0 \mod N \right\} \tag{2.15}$$

We say that any subgroup of $\mathrm{SL}_2(\mathbf{Z})$ which contains $\Gamma(N)$ for some N is a *congruence subgroup*; clearly $\Gamma_0(1) = \Gamma_1(1) = \Gamma(1) = \mathrm{SL}_2(\mathbf{Z})$. We call $\Gamma(N)$ the *principal congruence subgroup of level* N. If $\Gamma(N)$ is the largest principal congruence subgroup contained within a congruence subgroup N, then we say that Γ *has level* N. For instance, $\mathrm{SL}_2(\mathbf{Z})$ has level 1, and $\Gamma_0(2)$ has level 2.

We see that the following chain of inclusions holds, for any $N \in \mathbf{N}$:

$$\Gamma(N) \subseteq \Gamma_1(N) \subseteq \Gamma_0(N) \subseteq \mathrm{SL}_2(\mathbf{Z}). \tag{2.16}$$

We will call the congruence subgroups in this chain the *standard congruence subgroups.*

Occasionally in the literature we see other congruence subgroups; for instance, $\Gamma^0(N)$, which is the subset of $\mathrm{SL}_2(\mathbf{Z})$ with the *upper* right entry congruent to 0 modulo N. One downside of this particular group is that it does not contain $\left(\begin{smallmatrix} 1 & 1 \\ 0 & 1 \end{smallmatrix}\right)$ if $N > 1$, so modular forms for this group need not have a Fourier expansion in terms of q; we will consider this issue later.

We will often call an unspecified congruence subgroup Γ; this follows the notation in some of the literature, but we note that in [Koblitz (1993)] the group $\mathrm{SL}_2(\mathbf{Z})$ is called Γ. A natural further convention is to use γ for an element of Γ.

We will now show that the index $[\mathrm{SL}_2(\mathbf{Z}) : \Gamma(N)]$ is finite[1], compute the indices of these standard congruence subgroups inside $\mathrm{SL}_2(\mathbf{Z})$, and find coset representatives for these congruence subgroups in $\mathrm{SL}_2(\mathbf{Z})$. This will allow us, amongst other things, to find fundamental domains in $\mathcal{H}$ for these congruence subgroups.

We will need some basic theory of projective spaces over rings to find the index of our congruence subgroups within $\mathrm{SL}_2(\mathbf{Z})$.

Definition 2.10. Let N be a positive integer. We define

$$P(\mathbf{Z}/N\mathbf{Z}) := \{(a, b) \in (\mathbf{Z}/N\mathbf{Z})^2 : \exists c, d \in \mathbf{Z}/N\mathbf{Z} \text{ such that } ad - bc = 1\}.$$

The group of units $(\mathbf{Z}/N\mathbf{Z})^\times \subset (\mathbf{Z}/N\mathbf{Z})$ acts on $P(\mathbf{Z}/N\mathbf{Z})$ by $\lambda(a, b) = (\lambda a, \lambda b)$. We define

$$\mathbf{P}^1(\mathbf{Z}/N\mathbf{Z}) := P(\mathbf{Z}/N\mathbf{Z})/\sim,$$

where the equivalence relation $\sim$ is given by the action of $(\mathbf{Z}/N\mathbf{Z})^\times$. We write the equivalence class containing (a, b) as $(a : b)$.

[1] This will also show that all of the standard congruence subgroups have finite index.

If R is a field, $\mathbf{P}^1(R) \cong R \cup \{\infty\}$, as the equivalence classes are $(r : 1)$, for every element $r \in R$, plus the equivalence class $(1 : 0)$, which is called the *point at infinity*; this construction may be familiar to those readers who have encountered the theory of elliptic curves before. The structure of $\mathbf{P}^1(R)$ becomes more interesting when R is not a field, so there are more nonunits.

Proposition 2.11 (Size of $\mathbf{P}^1(\mathbf{Z}/N\mathbf{Z})$). *Let $N \in \mathbf{N}$. Then*

$$\#\mathbf{P}^1(\mathbf{Z}/N\mathbf{Z}) = N \prod_{p|N} (1 + 1/p).$$

Proof. We leave this as Exercise 9. □

Having set up this theory, we can now apply the preceding proposition to find indices of some of our congruence subgroups, and also to find explicit sets of coset representatives.

Proposition 2.12. *Let N be a positive integer. Then*

$$[\mathrm{SL}_2(\mathbf{Z}) : \Gamma_0(N)] = N \prod_{p|N} (1 + 1/p).$$

Proof. We define a map π from the quotient group $\Gamma_0(N)/\mathrm{SL}_2(\mathbf{Z})$ to $\mathbf{P}^1(\mathbf{Z}/N\mathbf{Z})$ by $\pi\left(\Gamma_0(N)\left(\begin{smallmatrix} a & b \\ c & d \end{smallmatrix}\right)\right) = (c : d)$. We must show that this map is well-defined, and then that it is a bijection. We will then use Proposition 2.11 to compute the dimension of the projective space.

Let $\Gamma_0(N)\left(\begin{smallmatrix} a & b \\ c & d \end{smallmatrix}\right)$ be a coset. We can change the matrix by left multiplication by $\left(\begin{smallmatrix} A & B \\ C & D \end{smallmatrix}\right) \in \Gamma_0(N)$ to obtain a matrix with bottom row $(aC + cD\ bC + dD)$. This acts as $(cD\ dD)$ on $\mathbf{P}^1(\mathbf{Z}/N\mathbf{Z})$ because $C \mod N \equiv 0$, and D is a unit in $\mathbf{Z}/N\mathbf{Z}$, so changing the coset representative doesn't change the action. This means that it is well-defined.

Now we show that it is a bijection. It is a surjection because given an element $\left(\begin{smallmatrix} a & b \\ c & d \end{smallmatrix}\right)$ of $\mathrm{SL}_2(\mathbf{Z})$ we have an element $(c : d)$ of $\mathbf{P}^1(\mathbf{Z}/N\mathbf{Z})$, and every element of $\mathbf{P}^1(\mathbf{Z}/N\mathbf{Z})$ appears in this way, as we can lift $(c : d)$ to an element of $\mathrm{SL}_2(\mathbf{Z})$ by lifting c and d from $\mathbf{Z}/N\mathbf{Z}$ to $\mathbf{Z}$ and then using Euclid's algorithm to find a suitable pair a and b.

Now we must show that π is injective; that is, if $\pi(\Gamma_0(N)\left(\begin{smallmatrix} a & b \\ c & d \end{smallmatrix}\right)) = \pi(\Gamma_0(N)\left(\begin{smallmatrix} A & B \\ C & D \end{smallmatrix}\right))$ then $\Gamma_0(N)\left(\begin{smallmatrix} a & b \\ c & d \end{smallmatrix}\right) = \Gamma_0(N)\left(\begin{smallmatrix} A & B \\ C & D \end{smallmatrix}\right)$, or in other words that $\left(\begin{smallmatrix} A & B \\ C & D \end{smallmatrix}\right) \cdot \left(\begin{smallmatrix} a & b \\ c & d \end{smallmatrix}\right)^{-1} \in \Gamma_0(N)$; the unique restriction here is that the lower left entry of the product, $Cd - cD$, must be congruent to 0 modulo N. The equality in $\mathbf{P}^1(\mathbf{Z}/N\mathbf{Z})$ is $(c : d) = (C : D)$; this means that there exists $\lambda \in (\mathbf{Z}/N\mathbf{Z})^\times$ such that $c = \lambda C$ and $d = \lambda D$. These two conditions are equivalent to one another, so π is injective. □

Finally, we will find the index of $\Gamma(N)$ in $\Gamma_0(N)$.

Proposition 2.13. *Let N be a positive integer. Then*

$$[\Gamma_0(N) : \Gamma(N)] = N^2 \prod_{p|N} \left(1 - \frac{1}{p}\right).$$

Proof. We can write the cosets as $\Gamma(N) \left(\begin{smallmatrix} a & b \\ 0 & d \end{smallmatrix}\right)$, with $a, b, d \in \mathbf{Z}/N\mathbf{Z}$. Each of these choices can be shown to be distinct, and there are N possibilities for b and $\phi(N)$ possibilities for a, and one choice of d given a. If we expand out the definition of $\phi(N)$ we will get the index given in the proposition, as required. □

As an example of this, let us consider $\Gamma_0(8)$, which has index $8 \cdot (1+1/2) = 12$ in $\mathrm{SL}_2(\mathbf{Z})$. We can write

$$\mathbf{P}^1(\mathbf{Z}/8\mathbf{Z}) = \{(n : 1)\}_{n \in [0...7]} \cup \{(1 : 2), (3 : 2), (5 : 2), (7 : 2)\};$$

(checking that these are all distinct is left as Exercise 10). From these, we can find a set of coset representatives using the lifting procedure detailed above; here is one such:

$$\left\{\begin{pmatrix} 1 & 0 \\ n & 1 \end{pmatrix}\right\}_{n \in [0...7]} \cup \left\{\begin{pmatrix} 1 & 1 \\ 1 & 2 \end{pmatrix}, \begin{pmatrix} -1 & -1 \\ 3 & 2 \end{pmatrix}, \begin{pmatrix} 3 & 1 \\ 5 & 2 \end{pmatrix}, \begin{pmatrix} 4 & 1 \\ 7 & 2 \end{pmatrix}\right\}.$$

We note that there will of course be many choices of coset representatives.

It should also be noted that we have shown (by combining Proposition 2.12 and Proposition 2.13) that every congruence subgroup has finite index in $\mathrm{SL}_2(\mathbf{Z})$. The converse to this is false; in [Hsu (1996)], for instance, an explicit example of a non-congruence subgroup G of index 10 is given (a non-congruence subgroup, for us, is a subgroup of $\mathrm{SL}_2(\mathbf{Z})$ of finite index which is not a congruence subgroup for any N). Using the SAGE package "Farey Symbol Functions" [Kurth (2007)], one can compute a set of generators for this non-congruence subgroup; for example, these matrices generate G:

$$\left\{\begin{pmatrix} 1 & 2 \\ 0 & 1 \end{pmatrix}, \begin{pmatrix} 2 & -1 \\ 7 & -3 \end{pmatrix}, \begin{pmatrix} 4 & -3 \\ 3 & -2 \end{pmatrix}\right\}.$$

In fact, there is a well-defined sense in which most subgroups of $\mathrm{SL}_2(\mathbf{Z})$ of finite index are non-congruence subgroups; see [Jones (1986)] for a discussion of this (one must be careful to make this statement precise, because there are countably many of both congruence and non-congruence subgroups).

2.5 Fundamental domains

We recall that a modular form satisfies the transformation equation given in (2.1). In this section, we will see that this means that we can determine the complex values that a modular form takes by considering its values on a certain subset of $\mathcal{H}$, called a fundamental domain.

We define a *fundamental domain* for a congruence subgroup Γ.

Definition 2.14. Let Γ be a congruence subgroup. A *fundamental domain F for* Γ is a closed subset of $\mathcal{H}$ such that

(1) every $z \in \mathcal{H}$ is Γ-equivalent to a point of the closure of F,
(2) no two distinct points in the interior of F are Γ-equivalent.

We note here that a fundamental domain F need not be a region, as it does not have to be connected. Also, one can remove closed sets without interiors from F and still have a fundamental domain; for instance, a closed line segment has no interior, so can be removed. We note that it is usual, however, to choose fundamental domains to be regions.

The standard example[2] of a fundamental domain is the region of $\mathcal{H}$ given by

$$F := \left\{ z : z \in \mathcal{H} \text{ and } |z| \geq 1 \text{ and } |\Re(z)| \leq \frac{1}{2} \right\}, \tag{2.17}$$

shown in Figure 2.1, which can be shown to be a fundamental domain for $\mathrm{SL}_2(\mathbf{Z})$ (see Exercise 7). We see also that any translate γF of F by an element γ of $\mathrm{SL}_2(\mathbf{Z})$ is also a fundamental domain for $\mathrm{SL}_2(\mathbf{Z})$, because every point of γF is equivalent to a point of F, and any two points of γF which are equivalent are not in the interior of γF. We also note that if z is a boundary (respectively, interior) point of F then γz is also a boundary (interior) point of γF because the Möbius transformation γ sends lines to lines. It is important to note that there are points on the boundary of F which are equivalent to other points on the boundary of F.

Once we have shown that the region F illustrated in Figure 2.1 is a fundamental domain for $\mathrm{SL}_2(\mathbf{Z})$, then we will be able to find fundamental domains for other congruence subgroups, using the following standard result:

[2] The Wikipedia article on fundamental domains [Wikipedia] says that "[t]his famous diagram [(Figure 2.1)] appears in all classical books on elliptic modular functions". This is likely to be true.

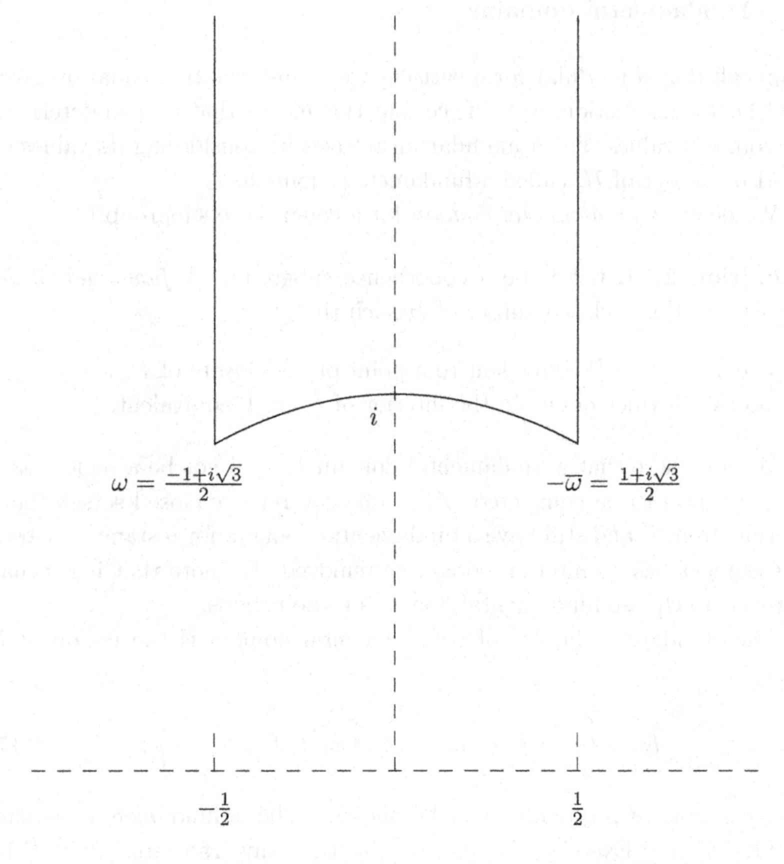

Fig. 2.1 The standard fundamental domain for $\mathrm{SL}_2(\mathbf{Z})$

Proposition 2.15. *Let Γ be a congruence subgroup of $\mathrm{SL}_2(\mathbf{Z})$, written as a disjoint union of cosets*

$$\mathrm{SL}_2(\mathbf{Z}) = \coprod_{i=1}^{n} \alpha_i \Gamma.$$

Then $G := \coprod_{i=1}^{n} \alpha_i^{-1} F$ is a fundamental domain for Γ.

Proof. We need to check the two conditions for being a fundamental domain; first we verify that every element z of $\mathcal{H}$ is Γ-equivalent to an element of G. Because F is a fundamental domain for $\mathrm{SL}_2(\mathbf{Z})$, we can find some $\delta \in \mathrm{SL}_2(\mathbf{Z})$ such that $\delta z \in F$. Now we have just written $\mathrm{SL}_2(\mathbf{Z})$ as a union of cosets, so there exist α_i and $\gamma \in \Gamma$ such that $\delta = \alpha_i \gamma$. This means that $\gamma z \in \alpha_i^{-1} F$, which was what we needed to show.

Now we will show that if z_1 and z_2 are two points in the interior of G, then they are not Γ-equivalent. We shall prove this by contradiction. Suppose that there exists $\gamma \in \Gamma$ such that $\gamma z_1 = z_2$. We can write $z_1 = \alpha_i^{-1} y_1$ and $z_2 = \alpha_j^{-1} y_2$ for some $y_i \in F$, and therefore we have the relation $(\alpha_j \gamma \alpha_i^{-1}) y_1 = y_2$. Now this is an $\mathrm{SL}_2(\mathbf{Z})$-equivalence between two points in the interior of F, but F is a fundamental domain, so there can be no such relation, so we have a contradiction, and therefore z_1 and z_2 are not Γ-equivalent. □

We now define the cusps of a congruence subgroup Γ. We define $\overline{\mathcal{H}}$ to be $\mathcal{H} \cup \{\infty\} \cup \mathbf{Q}$; we are adding a "point at infinity" ([Koblitz (1993)], Section III.1, suggests that this should be viewed as being a long way up the positive imaginary axis; it is sometimes called $i\infty$) and all of the rational numbers on the real axis. These added points $\{\infty\} \cup \mathbf{Q}$ are partitioned into equivalence classes by the action of Γ, which are called the *cusps*.

The standard abuse of notation here, which we shall follow, is to take one representative from each equivalence class and call this set *the cusps of* Γ.

If $\Gamma = \mathrm{SL}_2(\mathbf{Z})$ then there is a unique cusp, because if a/c is a fraction in lowest terms then we can use Euclid's algorithm to find integers b and d such that $\left(\begin{smallmatrix} a & b \\ c & d \end{smallmatrix}\right) \in \mathrm{SL}_2(\mathbf{Z})$; this matrix maps ∞ to a/c, so therefore every element of $\mathbf{Q}$ is equivalent to ∞. In general, there is more than one cusp; for instance, see Exercise 6.

As an illustration of this, we present a fundamental domain for $\Gamma_0(2)$ as Figure 2.2. This is based on choosing the coset representatives to be $\{\left(\begin{smallmatrix} 1 & 0 \\ 0 & 1 \end{smallmatrix}\right), \left(\begin{smallmatrix} 0 & -1 \\ 1 & 0 \end{smallmatrix}\right), \left(\begin{smallmatrix} 0 & -1 \\ 1 & 1 \end{smallmatrix}\right)\}$. We have chosen this to be connected, as is normal in the literature. From Exercise 6, we see that the cusps for $\Gamma_0(2)$ can be taken to be 0 and ∞, and if we look at the shape of the fundamental domain near 0 then we can see why the word "cusp" was chosen; it looks like a cusp.

It is important to reiterate the fact that there is no one canonical choice of fundamental domain for a given congruence subgroup. We can also choose a different fundamental domain for $\mathrm{SL}_2(\mathbf{Z})$ which does have a visible cusp; in Exercise 8 we will exhibit such a domain.

If Γ is a congruence subgroup, it is quite possible that a fundamental domain for Γ is not connected, although it is customary (for both computational and aesthetic reasons) to choose a fundamental domain so that it is connected.

If we know the values of a modular form f for Γ on a fundamental

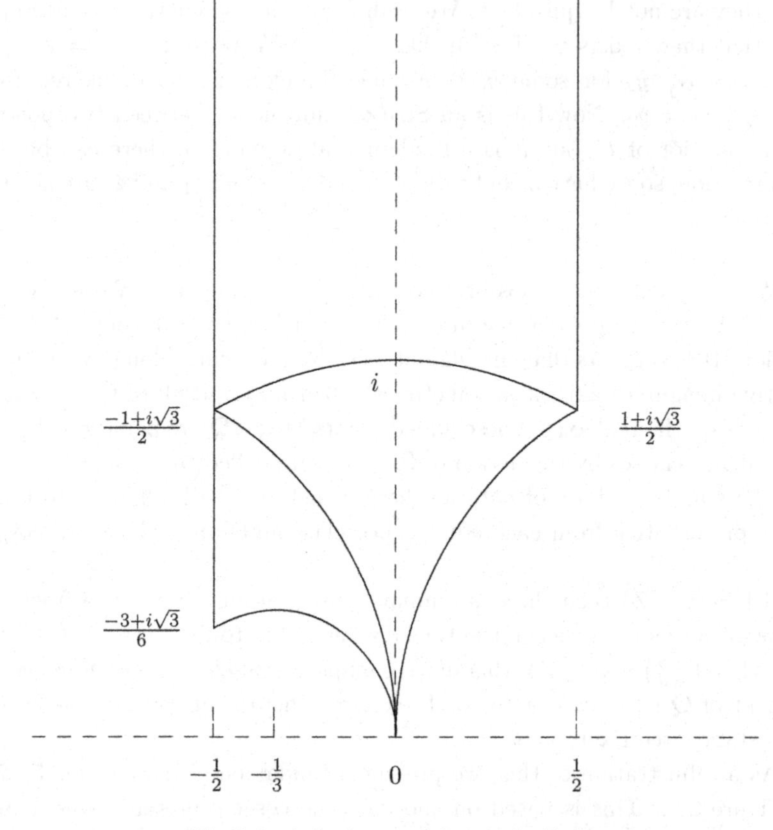

Fig. 2.2 A fundamental domain for $\Gamma_0(2)$

domain F for Γ, then we know its value for all of $\mathcal{H}$, because every point of $\mathcal{H}$ is Γ-equivalent to a point of F. This will allow us to find the dimensions of spaces of modular forms; we will show in Section 3.1 that a given nonzero modular form only has finitely many zeroes in its fundamental domain, and performing a contour integral around a carefully-chosen region of the fundamental domain will give us a dimension formula.

2.6 Modular forms for congruence subgroups

The full modular group is a natural choice for the first definition of modular forms, but there are objects which arise "in nature" which have the same type of behaviour as the modular forms we have already seen, but do not transform correctly under every element of $\mathrm{SL}_2(\mathbf{Z})$. Examples that we will consider in later chapters are the following series and products:

$$\theta(q) = \sum_{x,y\in\mathbf{Z}} q^{x^2+y^2},$$

and

$$f(q) = q\prod_{n=1}^{\infty}(1-q^n)^2(1-q^{11n})^2,$$

and

$$F(q) = \sum_{n>0 \text{ odd}} \sigma_1(n)q^n,$$

and

$$g(q) = q\prod_{n=1}^{\infty}(1-q^n)(1-q^{23n}).$$

The coefficients of all of these series are multiplicative, and the modular forms whose Fourier expansions these are all satisfy certain transformation properties, but they are not modular forms for $\mathrm{SL}_2(\mathbf{Z})$. We therefore broaden our definitions to include these forms.

Let f be a meromorphic function from $\mathcal{H}$ to $\mathbf{C}$, and let Γ be a fixed congruence subgroup. If we have that

$$f\left(\frac{az+b}{cz+d}\right) = (cz+d)^k \cdot f(z) \text{ for } \begin{pmatrix} a & b \\ c & d \end{pmatrix} \in \Gamma \text{ and } z \in \mathcal{H},$$

then we say that f *is weakly modular for* Γ.

We define the *weight* k *operator* $||[\gamma]_k$ on functions from $\mathcal{H}$ to $\mathbf{C}$ by

$$(f||[\gamma]_k)(z) = (cz+d)^{-k} f\left(\frac{az+b}{cz+d}\right), \text{ for } z \in \mathcal{H}. \tag{2.18}$$

We note for reference that there are many possible choices of notation for this operator; it has been written $f^{[\gamma]_k}$, $f|_k[\gamma]$, and $f[\gamma]_k$ by various authors.

We will now define modular forms for arbitrary congruence subgroups of $\mathrm{SL}_2(\mathbf{Z})$.

Definition 2.16 (Modular forms for congruence subgroups). *Let* Γ *be a congruence subgroup of* $\mathrm{SL}_2(\mathbf{Z})$ *and let* k *be an integer. We say that a function* $f : \mathcal{H} \to \mathbf{C}$ *is a* modular form of weight k for Γ *if*

(1) f *is holomorphic on* $\mathcal{H}$,

(2) f is weakly modular for Γ,
(3) $f\|[\alpha]_k$ is holomorphic at ∞ for all $\alpha \in \mathrm{SL}_2(\mathbf{Z})$.

If we also have that a_0 is 0 in the Fourier expansion of $f\|[\alpha]_k$ for every $\alpha \in \mathrm{SL}_2(\mathbf{Z})$, then we say that f is a cusp form of weight k for Γ. *If Γ contains $\Gamma(N)$, then we say that f is a modular form* of level N *or of level Γ. We write the space of modular forms of weight k for a congruence subgroup Γ as $M_k(\Gamma)$, and the space of cusp forms as $S_k(\Gamma)$.*

We note here that there is a theory of modular forms for non-congruence subgroups, but that this is outside the scope of this book. See [Li *et al.* (2005)] for a survey of recent results in this area.

We now make some remarks about the Fourier expansions of modular forms for congruence subgroups; this subject is more complicated than the $\mathrm{SL}_2(\mathbf{Z})$ case.

Remark 2.17. We saw that for modular forms for the full modular group $\mathrm{SL}_2(\mathbf{Z})$ there was a simple definition for a cusp form; we merely had to check that its Fourier expansion had no constant term. This is not sufficient in general for modular forms for smaller congruence subgroups; we have to check that it vanishes "at all the cusps". In Exercise 10 of Chapter 4 the reader is invited to find some non-cuspidal modular forms which do not have a constant term in their Fourier expansion at ∞.

Remark 2.18. We saw that modular forms for $\mathrm{SL}_2(\mathbf{Z})$ had Fourier expansions in terms of a variable $q = e^{2\pi i z}$, because we have that $f(z) = f(z+1)$. This still holds for the congruence subgroups $\Gamma_0(N)$ and $\Gamma_1(N)$, but we see that $\Gamma(N)$ does *not* contain $\left(\begin{smallmatrix}1&1\\0&1\end{smallmatrix}\right)$ if $N \geq 2$, so a modular form for $\Gamma(N)$ does not necessarily have a Fourier expansion in terms of q. Instead, it has a Fourier expansion in terms of $q^{1/N}$; some authors define $q_N := e^{(2\pi i z)/N}$.

We also have the notion of *modular forms with character.* This allows us to break up the space $S_k(\Gamma_1(N))$ into a direct sum of smaller spaces, and also gives a natural way of handling cases where a modular form does not quite transform in the standard way, but does transform in a slightly more general way.

Definition 2.19. Let N be a positive integer and let $\chi : (\mathbf{Z}/N\mathbf{Z})^\times \to \mathbf{C}^\times$ be a Dirichlet character (so χ is multiplicative and $\chi(1) = 1$). We say that a function $f : \mathcal{H} \to \mathbf{C}$ which is holomorphic on $\mathcal{H}$ and is holomorphic at the cusps of $\Gamma_1(N)$ is a *modular form of weight k for $\Gamma_0(N)$ with character χ* if

$$f||[\gamma]_k = \chi(d)f \text{ for all } \left(\begin{smallmatrix} a & b \\ c & d \end{smallmatrix}\right) \in \Gamma_0(N).$$

We write the space of modular forms of weight k and character χ for a congruence subgroup Γ as $M_k(\Gamma, \chi)$, and the space of cusp forms of weight k and character χ as $S_k(\Gamma, \chi)$.

We note that if $\chi(-1) \neq (-1)^k$, then $M_k(\Gamma_0(N), \chi) = 0$; this follows because $\left(\begin{smallmatrix} -1 & 0 \\ 0 & -1 \end{smallmatrix}\right) \in \Gamma_0(N)$, and if we substitute this into the formula 2.18 for the weight k operator, we obtain

$$f||[\left(\begin{smallmatrix} -1 & 0 \\ 0 & -1 \end{smallmatrix}\right)]_k = (-1)^k f = \chi(-1)f,$$

and this means that f must be identically zero.

We will now show that we can transfer the study of forms on $\Gamma(N)$ to the study of forms on $\Gamma_1(N^2)$, by conjugation; this follows the approach taken in [Lang (1976)], Section III.2. We take the matrix

$$\alpha := \begin{pmatrix} N & 0 \\ 0 & 1 \end{pmatrix},$$

and note that the map $\gamma \mapsto \alpha\gamma\alpha^{-1}$ is an inner automorphism of $\mathrm{GL}_2(\mathbf{Q})$. We will call this map $\star$, following Lang's notation. We see that if $\gamma = \left(\begin{smallmatrix} a & b \\ c & d \end{smallmatrix}\right) \in \Gamma$, then

$$\gamma^\star = \begin{pmatrix} a & b/N \\ cN & d \end{pmatrix},$$

and that $\gamma^\star \in \Gamma_0(N^2)$ (because $cN \equiv 0 \mod N^2$) and $\gamma^\star \in \Gamma_1(N)$. This means that the group $\Gamma^\star(N) := \alpha\Gamma(N)\alpha^{-1}$ contains $\Gamma_1(N^2)$ and is a subgroup of $\mathrm{SL}_2(\mathbf{Z})$, so is a congruence subgroup. Therefore, if f is a modular form for $\Gamma^\star(N)$ then it is also a modular form for $\Gamma_1(N^2)$. We see that if $f \in M_k(\Gamma(N))$, then $f||[\alpha]_k \in M_k(\Gamma^\star(N))$, because

$$(f||[\alpha]_k)\,||[\alpha^{-1}\gamma\alpha]_k = f||[\alpha\alpha^{-1}\gamma\alpha]_k = f||[\alpha]_k;$$

this follows because we have $f||[\alpha\beta]_k = (f||[\alpha]_k)||[\beta]_k$ (see Exercise 19), and f is a modular form for $\Gamma(N)$. We can check that it is holomorphic at the cusps in a similar way. We also note that if f is a modular form for Γ then it is also a modular form for congruence subgroups contained within Γ. We have proved that

Proposition 2.20. *If we have a modular form for $\Gamma(N)$, we can transform it into a modular form for $\Gamma_1(N^2)$.*

This is important because the transformed modular form $f^\star = f||[\alpha]_k$ for $\Gamma_1(N^2)$ will have a Fourier expansion in terms of q, which a modular form f for $\Gamma(N)$ may not.

2.7 Eisenstein series for congruence subgroups

It is intuitive that there should be modular forms for proper congruence subgroups Γ which are not modular forms for $\mathrm{SL}_2(\mathbf{Z})$; for a holomorphic function $f : \mathcal{H} \to \mathbf{C}$, being a modular form for Γ is a less strict condition than being a modular form for $\mathrm{SL}_2(\mathbf{Z})$, as it means that f has to satisfy fewer conditions. We will make this intuition precise in this section.

Let N be an integer; in this section, following the presentation given in [Koblitz (1993)], Section III.3, we will explicitly construct Eisenstein series for $\Gamma(N)$ and $\Gamma_1(N)$. Also, we will explicitly construct modular forms of *odd* weight k.

We let $\mathbf{a} := (a_1, a_2)$ be a pair of integers modulo N, and let k be an integer which is at least 3 (note that k can be odd or even). We define the following function, which, following Koblitz, we call the *level N Eisenstein series of weight k*:

$$G_k^{\mathbf{a}}(z) := G_k^{\mathbf{a} \mod N}(z) = \sum_{\substack{\mathbf{m}\in\mathbf{Z}^2 \\ \mathbf{m}\equiv\mathbf{a} \mod N}} \frac{1}{(m_1 z + m_2)^k}.$$

We note that $G_k^{(0,0)}(z) = N^{-k} \cdot G_k(z)$, which is the Eisenstein series of level 1 that we discussed earlier, because

$$G_k^{(0,0)}(z) = N^{-k} \cdot \sum_{m_1, m_2 \in \mathbf{Z}} \frac{1}{(m_1 z + m_2)^k} = N^{-k} G_k(z).$$

We will therefore assume that $\mathbf{a} \neq (0,0)$ in this section.

Theorem 2.21. *Let $N \geq 1$ and $k \geq 3$ be integers, and let $\mathbf{a} = (a_1, a_2)$. Then*

$$G_k^{\mathbf{a} \mod N}(z) \in M_k(\Gamma(N)), \text{ and } G_k^{(0,a_2) \mod N}(z) \in M_k(\Gamma_1(N)).$$

We note that there are going to be essentially as many distinct Eisenstein series as there are cusps, because if we have two pairs (a_1, a_2) and (b_1, b_2) which are equivalent modulo $\Gamma_1(N)$, then the Eisenstein series we obtain will be the same up to the weight k action of $\Gamma_1(N)$.

Proof. In this proof we will let Γ be either $\Gamma(N)$ or $\Gamma_1(N)$, depending on whether $\mathbf{a} = (a_1, a_2)$ or $(0, a_2)$.

Firstly, we must check that $G_k^{\mathbf{a} \mod N}(z)$ is holomorphic on $\mathcal{H}$. This is where we use the assumption that $k \geq 3$; using Lemma 2.7, we see that the series is absolutely and uniformly convergent on $\mathcal{H}$. If $k = 1$ or $k = 2$ then this would *not* hold, as the series fails to converge absolutely.

Now we will show that $G_k^{\mathbf{a} \bmod N}(z)$ transforms correctly under the action of elements of Γ. We will use a similar method to that used in the proof that the level 1 Eisenstein series $G_k(z)$ is a modular form of weight k. We take $\gamma = \left(\begin{smallmatrix} a & b \\ c & d \end{smallmatrix}\right) \in \Gamma$; this means that $a \equiv d \equiv 1 \mod N$ and $c \equiv 0 \mod N$. By the definitions, we have

$$G_k^{\mathbf{a} \bmod N}(z)|[\gamma]_k \tag{2.19}$$

$$= (cz+d)^{-k} \cdot \sum_{\mathbf{m} \equiv \mathbf{a} \bmod N} \left(m_1 \cdot \frac{az+b}{cz+d} + m_2 \right)^{-k} \tag{2.20}$$

$$= \sum_{\mathbf{m} \equiv \mathbf{a} \bmod N} \frac{1}{((m_1 a + m_2 c)z + (m_1 b + m_2 d))^k}. \tag{2.21}$$

We define $\mathbf{m}' = (m_1 a + m_2 c, m_1 b + m_2 d) = (m_1 m_2)\left(\begin{smallmatrix} a & b \\ c & d \end{smallmatrix}\right) = \mathbf{m}\gamma$. Because $\gamma \in \Gamma_1(N)$ then we have that $\mathbf{m}'$ is congruent to $\mathbf{a}\gamma$ modulo N. This means that we can rearrange the order of the summation, and we find that, because the action of $\left(\begin{smallmatrix} a & b \\ c & d \end{smallmatrix}\right)$ gives a one-to-one mapping, we have that $G_k^{\mathbf{a} \bmod N}$ is invariant under the action of γ, which is what we needed to show.

Finally, one must check that $G_k^{\mathbf{a} \bmod N}(z)$ does not have poles at the cusps of Γ. By analogy to the previous paragraph's result, we see that the action of $\mathrm{SL}_2(\mathbf{Z})$ permutes the Eisenstein series $G_k^{\mathbf{a} \bmod N}$, so it is enough to show that each of these series is holomorphic at ∞.

This is very similar to the proof that the Eisenstein series G_k for $\mathrm{SL}_2(\mathbf{Z})$ is holomorphic at ∞; the limit as $z \to i\infty$ of $G_k^{\mathbf{a} \bmod N}$ is strictly less than $2\zeta(k)$, because not all of the terms that contribute in the level 1 case will contribute here. In particular, it is finite and well-defined, because we have assumed that the weight k is at least 3. □

Remark 2.22. If we have $\Gamma = \Gamma(N)$ and $N \geq 3$ then $\left(\begin{smallmatrix} 1 & 1 \\ 0 & 1 \end{smallmatrix}\right) \notin \Gamma$, so the modular form $G_k^{\mathbf{a} \bmod N}$ is not automatically identically zero. We will see from the explicit calculations of Theorem 2.23 that these modular forms are in general not zero, so we will have constructed explicit modular forms of odd weight.

We can compute the Fourier coefficients of $G_k^{\mathbf{a} \bmod N}(z)$ explicitly, by using a similar proof to that for the $\mathrm{SL}_2(\mathbf{Z})$ case. We will not give the general result or its proof here, but will instead state two important special cases of this result.

Theorem 2.23. *Let $k \geq 3$ and $N \geq 1$ be positive integers, and let* $\mathbf{a} := (a_1, a_2) \in \mathbf{Z}^2$. *We define* $q_N := e^{(2\pi i z)/N}$ *and* $\zeta = e^{2\pi i/N}$.

If $\mathbf{a} = (a_1, 0)$, *then we have that*

$$G_k^{(a_1,0) \mod N}(z) = b_{0,k}^{(a_1,0)} + \sum_{n=1}^{\infty} b_{n,k}^{(a_1,0)} \cdot q_N^n,$$

where

$$b_{n,k}^{(a_1,0)} := c_k \cdot \left(\sum_{j|n} j^{k-1} + (-1)^k \sum_{j|n} j^{k-1} \right),$$

and

$$c_k := \frac{(-1)^{k-1} 2k\zeta(k)}{N^k B_k}.$$

If $\mathbf{a} = (0, a_2)$, *then we have that*

$$G_k^{(0,a_2)}(z) = b_{0,k}^{(0,a_2)} + c_k \sum_{n=1}^{\infty} \left(\sum_{j|n} j^{k-1}(\zeta^{j \cdot a_2} + (-1)^k \cdot \zeta^{-j \cdot a_2}) \right) \cdot q^n$$

where

$$b_{n,k}^{(0,a_2)} = 0 if N \nmid n \text{ and } b_{Nn,k}^{(0,a_2)} = c_k \sum_{j|n} j^{k-1} \left(\zeta^{ja_2} + (-1)^k \zeta^{-ja_2} \right).$$

Remark 2.24. We see here that $G_k^{(0,a_2)}(z)$, which is a modular form for $\Gamma_1(N)$, has a Fourier expansion in terms of q, whereas $G_k^{(a_1,0)}(z)$, which is a modular form for the smaller congruence subgroup $\Gamma(N)$, does not. This is because $\left(\begin{smallmatrix} 1 & 1 \\ 0 & 1 \end{smallmatrix}\right) \in \Gamma_1(N)$, but $\left(\begin{smallmatrix} 1 & 1 \\ 0 & 1 \end{smallmatrix}\right) \notin \Gamma(N)$, for $N > 1$.

Proof. The intricacies of this proof can be found in [Koblitz (1993)], as Proposition III.3.22; one splits the sum up into two parts, and considers generalizations of the formula relating the sum of reciprocals of $(z + n)$ and the Fourier expansion given in (2.8). Finally, one gathers up the terms in q^i for each i to obtain the formulae for the Fourier expansions given in the theorem. □

Remark 2.25. We notice here that we have constructed explicit nonzero examples of modular forms of odd weight, which shows explicitly that there are modular forms for congruence subgroups of $\mathrm{SL}_2(\mathbf{Z})$ which are not modular forms for $\mathrm{SL}_2(\mathbf{Z})$.

It should be noted also that one can construct modular forms of weights 1 and 2 for congruence subgroups, but that one has to use a modified approach, as convergence issues become more delicate; for instance, we can no longer use Lemma 2.7. One reference for this is Section 7.2 of [Miyake (2006)].

2.8 Derivatives of modular forms

Although this book deals mainly with number-theoretic questions, modular forms as we have defined them here are complex functions on $\mathcal{H}$, so it makes sense to talk about their derivatives. The complex nature of modular forms will recur in later chapters, most notably in the proof in Chapter 3 that spaces of modular forms of a given weight and level are finite-dimensional.

We will define a differential operator θ (also known as the *Ramanujan operator*) by its action on Fourier expansions:

$$\theta := q\frac{d}{dq} = \frac{1}{2\pi i} \cdot \frac{d}{dz}. \tag{2.22}$$

The second equality follows from the change of variable from q to z. We also define a weight k operator $\partial_k : M_k(\mathrm{SL}_2(\mathbf{Z})) \to M_{k+2}(\mathrm{SL}_2(\mathbf{Z}))$ by

$$\partial = \partial_k := 12\theta - k \cdot E_2. \tag{2.23}$$

This operator is interesting because we have the following theorem that tells us that ∂_k sends modular forms to modular forms.

Theorem 2.26. *Let k be a non-negative integer. The weight k operator ∂_k maps $M_k(\mathrm{SL}_2(\mathbf{Z}))$ into $M_{k+2}(\mathrm{SL}_2(\mathbf{Z}))$, and maps cusp forms to cusp forms.*

Proof. Let $f \in M_k(\mathrm{SL}_2(\mathbf{Z}))$ and let $g = 12\theta(f) - k \cdot E_2 f$. We will show that g is a modular form of weight $k+2$. We may assume that k is even here, as otherwise the result is trivial (0 is mapped to 0).

Firstly we must check that g is holomorphic at ∞ and on $\mathcal{H}$. We see that it is holomorphic at ∞ by checking that both $\theta(g)$ and $E_2 \cdot g$ have only non-negative Fourier coefficients. We leave holomorphicity on $\mathcal{H}$ as an exercise (Exercise 15).

We now check that it satisfies the transformation formulae, by looking at how f transforms under $\left(\begin{smallmatrix} 0 & 1 \\ -1 & 0 \end{smallmatrix}\right)$:

$$f\left(\frac{-1}{z}\right) = z^k f(z); \tag{2.24}$$

this comes from substituting $\left(\begin{smallmatrix} 0 & 1 \\ -1 & 0 \end{smallmatrix}\right)$ into the standard transformation formula. We differentiate (2.24) with respect to z to obtain

$$\begin{aligned} \frac{1}{z^2} f'\left(\frac{-1}{z}\right) &= z^k f'(z) + kz^{k-1} f(z) \\ \Rightarrow f'\left(\frac{-1}{z}\right) &= z^{k+2} f'(z) + kz^{k+1} f(z). \end{aligned}$$

If we can show that g transforms correctly under this and under $\left(\begin{smallmatrix}1&1\\0&1\end{smallmatrix}\right)$, then we will have proved that g transforms correctly under the generators of $\mathrm{SL}_2(\mathbf{Z})/\{\pm 1\}$ which is enough to prove our result; we note that because g has a Fourier expansion in q that it does transform under $z \mapsto z+1$. We now perform the calculation, using Proposition 2.9 to get the transformation formula for E_2.

$$\begin{aligned}
g(-1/z) &= \frac{1}{2\pi i} f'(-1/z) - \frac{k}{12} E_2(-1/z) f(-1/z) \\
&= \frac{1}{2\pi i}\left(z^{k+2} f'(z) + k z^{k+1} f(z)\right) - \frac{k}{12}\left(E_2(z) z^2 + \frac{12}{2\pi i z} z^2\right) z^k f(z) \\
&= z^{k+2}\left(\frac{1}{2\pi i} f'(z) - \frac{k}{12} E_2(z) f(z)\right) + \frac{1}{2\pi i} z^{k+1}\left(f(z) - f(z)\right) \\
&= z^{k+2}\left(\frac{1}{2\pi i} f'(z) - \frac{k}{12} E_2(z) f(z)\right) \\
&= z^{k+2} g(z),
\end{aligned}$$

which is what we needed to show.

To check that cusp forms are sent to cusp forms, we note that the Fourier expansion of $E_2 \cdot g$ has a nonzero constant term if and only if f has a nonzero constant term, and $\theta(g)$ has no constant term, so their sum has a nonzero constant term if and only if g has a nonzero constant term. We note that we are using the fact that $\mathrm{SL}_2(\mathbf{Z})$ has a unique cusp to make this statement; it will not necessarily be true in general. □

As an example of this, we note that

$$\partial_4 E_4 = -4E_6,\ \partial_6 E_6 = -6E_8 \text{ and } \partial_8 E_8 = -8E_{10}.$$

Because the dimensions of $M_4(\mathrm{SL}_2(\mathbf{Z}))$, $M_6(\mathrm{SL}_2(\mathbf{Z}))$ and $M_8(\mathrm{SL}_2(\mathbf{Z}))$ are all 1 (which we will prove in Chapter 3) we see that identities of this form must exist; the only question is what the constant terms are, and we find these by noticing that the constant term is given by $-k$, because $-kE_2 f$ has constant term $-k$ and $\theta(g)$ has no constant term.

It can also be seen that ∂_k acts on modular forms like a derivation; in other words, if g and h are modular forms of weight k and l respectively, then

$$\partial_{k+l}(g \cdot h) = \partial_k(g) \cdot h + g \cdot \partial_l(h).$$

This follows because

$$\begin{aligned}
\partial_{k+l}(g \cdot h) &= 12\theta(g \cdot h) - (k+l) E_2 \cdot g \cdot h \\
&= 12\theta(g) h - k E_2 \cdot g \cdot h + 12\theta(h) g - l E_2 \cdot g \cdot h \\
&= \partial_k(g) h + \partial_l(h) g.
\end{aligned}$$

Theorem 2.26 is given in the literature in several places; in [Koblitz (1993)], as Exercise III.2.7 and in [Lang (1976)], as Theorem X.5.3, for instance. It can be generalized naturally to the following result:

Theorem 2.27. *Let Γ be a congruence subgroup and let $f \in M_k(\Gamma)$. Then $\partial(f) \in M_{k+2}(\Gamma)$.*

Proof. This is a generalization of the theorem for $\mathrm{SL}_2(\mathbf{Z})$; this is left as an exercise for the reader, as Exercise 18. □

Finally, let us show that θ "destroys modularity"; in other words, that if f is a nonconstant modular form, then $\theta(f)$ is *not* a modular form.

Proposition 2.28. *Let f be a non-constant modular form of weight k for the congruence subgroup Γ. Then $\theta(f)$ is not a modular form.*

Proof. We use the definition of the δ_k operator in terms of E_2 and θ given in (2.22); we see that

$$\theta(f) = \frac{\partial_k(f) - k \cdot E_2 f}{12}. \tag{2.25}$$

Now a non-constant modular form can only have one weight (see Exercise 16), and Theorem 2.26 shows us that $\delta_k(f)$ has weight $k+2$. We see (from Proposition 2.9) that while $\delta_k(f)$ transforms correctly under the weight $k+2$ action, $E_2 f$ does not, so we see by substituting in a suitable element of Γ that $\theta(f)$ is not modular. □

The results of this section are all special cases of the work of Rankin, who proved results in [Rankin (1956)] which classify which polynomials of degree n in the first n derivatives of a modular form $f(z)$ are modular forms.

2.8.1 *Quasi-modular forms*

To quote Koblitz it "might seem unfortunate" that we have one Eisenstein series, E_2, in the above proceedings which is not a modular form, although it plays an important role in the definition of θ and ∂_k.

In [Kaneko and Zagier (1995)], a generalization of the notion of modular forms which allows us to include the Eisenstein series E_2 can be found. These forms are called the quasi-modular forms.

Definition 2.29 ([Kaneko and Zagier (1995)], Proposition 1).
Let Γ be a congruence subgroup. The ring of quasi-modular forms *for Γ*

are given by polynomials in E_2 with coefficients which are modular forms for Γ.

By explicit computation (see Exercise 17), we can see that $\theta(E_2)$ is a quasi-modular form, so this ring is closed under the action of θ. The quasi-modular forms play an important role in string theory and mirror symmetry (see [Bryan and Leung (2000)], for example); they also appear in number theory. The article [Atkin and Garvan (2003)], for instance, uses the theory of quasi-modular forms in its proof of results on partitions.

2.9 Exercises

(1)(a) Prove that $\mathrm{SL}_2(\mathbf{Z})$ is generated by the matrices S and T given in the text.

(b) Prove that the only matrices in $\mathrm{SL}_2(\mathbf{Z})$ which act trivially on $\mathcal{H}$ are $\pm\left(\begin{smallmatrix}1&0\\0&1\end{smallmatrix}\right)$.

(2) Let k be an integer and let Γ be a congruence subgroup. Prove that $M_k(\Gamma)$ and $S_k(\Gamma)$ are complex vector spaces and that $\mathcal{M}(\Gamma)$ and $\mathcal{S}(\Gamma)$ are graded rings. Prove that $\mathcal{S}(\Gamma)$ is an ideal in $\mathcal{M}(\Gamma)$.

(3) Let $k \geq 4$ be an even integer. Prove that the Fourier coefficients of E_k are multiplicative.

(4) Let $z \in \mathcal{H}$ and let $\gamma := \left(\begin{smallmatrix}a&b\\c&d\end{smallmatrix}\right) \in \mathrm{SL}_2(\mathbf{Z})$. Show that

$$\Im(\gamma(z)) = \frac{\Im(z)}{|cz+d|^2}.$$

Show that if $g(z) := |f(z)| \cdot y^{k/2}$, where $f \in M_k(\mathrm{SL}_2(\mathbf{Z}))$, then $g(z)$ is invariant under the action of $\left(\begin{smallmatrix}a&b\\c&d\end{smallmatrix}\right)$.

(5) Let $k \geq 4$ be an even integer. Show that $M_k(\mathrm{SL}_2(\mathbf{Z})) = \mathbf{C}E_k \oplus S_k(\mathrm{SL}_2(\mathbf{Z}))$:

(a) By considering the cuspidal condition, show that either $f \in M_k(\mathrm{SL}_2(\mathbf{Z}))$ is a cusp form, or that there exists $0 \neq c \in \mathbf{C}$ such that $f - c \cdot E_k$ is a cusp form.

(b) Check that this is a direct sum.

(6) Prove that $\Gamma_0(p)$ has two cusps, which we can take to be 0 and ∞, and find the three cusps for $\Gamma(2)$.

(7) Show that the region F of $\mathcal{H}$ defined in equation 2.17 is a fundamental domain for $\mathrm{SL}_2(\mathbf{Z})$.

(a) Show that every element of $\mathcal{H}$ is equivalent to an element z with $|\Re(z)| \leq 1/2$.
(b) We now need to show that we can take our element z in the strip $-1/2 \leq z \leq 1/2$ to have $|z| \geq 1$. Roughly speaking, we will use S to "bounce" z out of the central strip, then use T to translate it back, until z has large enough modulus.

i. Show that, for a fixed point $z \in \mathcal{H}$, $|cz + d|$ is bounded away from 0.
ii. Show that there exists $\gamma \in \mathrm{SL}_2(\mathbf{Z})$ such that $\Im(\gamma z)$ is maximal, and that we can take γz to have $|\Re(z)| \leq 1/2$.
iii. Note that if $\gamma z \notin F$, then we have $\Im(-1/(\gamma z)) > \Im(z)$, and use the contradiction here to prove the result.

(8) Find a fundamental domain for $\mathrm{SL}_2(\mathbf{Z})$ that has a visible cusp. You may wish to use [Verrill (2001)] to draw pictures of this to check that it really does have a visible cusp.
(9) Prove Proposition 2.11.
(10) Show that the presentation given of $\mathbf{P}^1(\mathbf{Z}/8\mathbf{Z})$ is really a presentation of this space. Check that the given coset representatives for $\Gamma_0(8)$ in $\mathrm{SL}_2(\mathbf{Z})$ are distinct.
(11) Let N be a positive integer. Find the index of $\Gamma_1(N)$ inside $\Gamma_0(N)$.
(12) Let N be a positive integer. Using the techniques of this chapter, find a set of coset representatives for $\Gamma_0(N)$ in $\mathrm{SL}_2(\mathbf{Z})$, and then find a fundamental domain for $\Gamma_0(N)$. Sketch this fundamental domain.
(13) In this exercise we will examine ways of constructing new congruence subgroups from old ones.

(a) Let Γ be a congruence subgroup of $\mathrm{SL}_2(\mathbf{Z})$ and let g be an element of $\mathrm{SL}_2(\mathbf{Z})$. Show that $g^{-1}\Gamma g$ is also a congruence subgroup.
(b) Let Γ_1 and Γ_2 be congruence subgroups. Show that $\Gamma_1 \cap \Gamma_2$ is also a congruence subgroup.
(c) Let $\alpha \in \mathrm{GL}_2(\mathbf{Q})$ and let Γ be a congruence subgroup. Show that $\alpha\Gamma\alpha^{-1} \cap \mathrm{SL}_2(\mathbf{Z})$ is a congruence subgroup. What happens if we simply consider $\alpha\Gamma\alpha^{-1}$; is this a congruence subgroup?

(14) Compute Δ'; the derivative of the Δ function with respect to z. Using this, compute $\partial_{12}\Delta$, and explain the result.
(15) Show that, if f is a modular form of weight k, then $\partial_k(f)$ is holomorphic on $\mathcal{H}$.
(16) Let Γ be a congruence subgroup and let f be a modular form for Γ. Show that if f has weight k and weight l for two distinct integers k

and l, then f must be a constant.

(17) By explicit computation, or otherwise, show that $(E_2^2 - 12\theta(E_2))(q)$ is the Fourier expansion of a modular form for $\mathrm{SL}_2(\mathbf{Z})$ of weight 4, and use this to prove that E_2 is quasimodular.

(18) Prove that, if f is a modular form of weight k for the congruence subgroup Γ, then $\partial_k(f)$ is a modular form of weight $k+2$ for the same congruence subgroup Γ.

(a) Generalize the proof that $\partial_k(f)$ transforms correctly under the action of elements γ of Γ; you may find it useful to recall that $\det \gamma = 1$ for all γ.

(b) If f is a cusp form of weight k for Γ, is $\partial_k(f)$ a cusp form? Consider what happens at *all* of the cusps, not just ∞.

(19) In this exercise, we will prove some results about the weight-k action on modular forms.

(a) Let Γ be a congruence subgroup, and let f be a modular form of weight k for Γ. If $\gamma_1, \gamma_2 \in \Gamma$, then show that

$$f|[\gamma_1\gamma_2]_k = (f|[\gamma_1]_k)|[\gamma_2]_k.$$

(b) We can generalize the weight-k action in the following way. Let k be a positive integer and let $\alpha = \left(\begin{smallmatrix} a & b \\ c & d \end{smallmatrix}\right) \in \mathrm{GL}_2^+(\mathbf{Q})$. We define the *generalized weight-k action* on a modular form f to be

$$(f|[\alpha]_k)(z) = (\det \alpha)^{k-1}(cz+d)^{-k} f\left(\frac{az+b}{cz+d}\right).$$

Show that $f|[\alpha\beta]_k = (f|[\alpha]_k)|[\beta]_k$ if $\alpha, \beta \in \mathrm{GL}_2^+(\mathbf{Q})$.

Chapter 3

Results on finite-dimensionality

[W]e should like it to be finite, like our mind. — Gottfried William Leibnitz.

We have already remarked that, although we are interested in modular forms for their number-theoretic properties, we will use their complex analytic nature to obtain useful results on the finite-dimensionality of spaces of modular forms. This chapter will give us the complex tools we will need to obtain our arithmetic results.

3.1 Spaces of modular forms are finite-dimensional

The titular result of this section is a great aid to computation; it means that one can consider only finite-dimensional vector spaces, which means that we can use computers to find spaces of modular forms and perform computations on them, knowing that computations in a finite-dimensional vector space with a given basis will terminate in a finite amount of time.

Finite-dimensionality is computationally very important; we will see later that spaces of modular *functions* — functions which transform like modular forms but are not necessarily holomorphic on $\mathcal{H}$ or at ∞ or other cusps — are not finite-dimensional, so many of the results that we will prove here do not carry over into the modular functions setting; we will consider this further in Section 5.1.

We now state the important result on spaces of modular forms for a given congruence subgroup which enables our computations.

Theorem 3.1. *Let Γ be a congruence subgroup, and let k be an integer. The spaces $S_k(\Gamma)$ and $M_k(\Gamma)$ are finite-dimensional complex vector spaces.*

To prove this theorem, we will first prove the following result concerning the zeroes of a modular form for the full modular group, which will allow us to show that, if k is an integer, then $M_k(\mathrm{SL}_2(\mathbf{Z}))$ is a finite-dimensional complex vector space.

We will then use finite-dimensionality for spaces of modular forms over $\mathrm{SL}_2(\mathbf{Z})$ to imply finite-dimensionality in general. We will also quote without proof some standard results on dimensions of spaces of modular forms for the standard congruence subgroups, which we introduced in Section 2.4.

Proposition 3.2. *Let $f(z)$ be a nonzero weakly modular form of weight k for $\mathrm{SL}_2(\mathbf{Z})$. Let P be an element of $\mathcal{H}$. We define $v_P(f)$ to be the order of zero (or minus the order of pole) at the point P, and we define $v_\infty(f)$ to be the index of the first non-vanishing term in the Fourier expansion of f. Then the following formula (known as the "residue formula" or the "valence formula") holds:*

$$v_\infty(f) + \frac{1}{2} \cdot v_i(f) + \frac{1}{3} \cdot v_\omega(f) + \sum_{\substack{P \in \mathrm{SL}_2(\mathbf{Z}) \backslash \mathcal{H} \\ P \neq i, \omega}} v_P(f) = \frac{k}{12}. \tag{3.1}$$

The following argument is standard, and appears in many texts on modular forms. It uses complex analysis, relying on the fact that modular forms are complex-analytic objects.

This sits somewhat uneasily within a number-theoretic context; it can seem odd that to prove results about number theory we must go through analysis, but there is a famous quote attributed to Hadamard[1] saying that "the shortest path between two truths in the real domain passes through the complex domain"; this may also be true with "real" replaced by "number-theoretic".

We note also that the argument we are about to give also gives a proof of a special case of the Riemann-Roch theorem in a concrete way; see [Hartshorne (1977)], Section IV.1, for more details on Riemann-Roch.

Proof. [Proof of Proposition 3.2] We will follow the exposition given in [Koblitz (1993)], Chapter III.2. The proof proceeds by performing an integral around the contour shown in Figure 3.1; we will call this contour $\mathcal{C}$. This is the so-called standard fundamental domain, which appeared as Figure 2.1.

[1] It is actually due to Painlevé.

Following [Serre (1973a)] and [Koblitz (1993)], we have added a point P on the vertical boundary (and therefore a corresponding point $T(P)$ which is also on the vertical boundary) and possible zeroes or poles at the points ω and $-\bar{\omega}$ and at i; we note that both i and ω appear separately in (3.1). The contour is drawn so as to include exactly one of P and $T(P)$, and to avoid i, ω and $-\overline{\omega}$. If there are several poles or zeroes on the boundary, then the integration would proceed in a similar way, as there are only finitely many zeroes or poles within $\mathcal{C}$.

Let us make the change of variables $q = e^{2\pi i z}$, so $\tilde{f}(q)$ is the q-expansion of $f(z)$ in a neighbourhood of the origin. We are being careful here to draw a distinction between the Laurent series $\tilde{f}$ of f and the function $f(z)$ itself.

The line at the top of the contour is the horizontal line from $A = \frac{1}{2} + Ri$ to $B = -\frac{1}{2} + Ri$, where R is chosen so that it is larger than the imaginary part of any of the zeroes or poles of $f(z)$. We can find such an R because $\tilde{f}$ is a *meromorphic* function, so there exists some positive real number s such that $\tilde{f}$ has neither zeroes or poles for $0 < |q| < s$; from this, it follows that f has no zeroes or poles if $\Im(z) > \frac{1}{2\pi} \log(1/s)$.

By the residue theorem from complex analysis, we have that

$$\frac{1}{2\pi i} \int_{\mathcal{C}} \frac{f'(z)}{f(z)} dz = \sum_{\substack{P \in \mathrm{SL}_2(\mathbf{Z}) \backslash \mathcal{H} \\ P \neq i, \omega}} v_P(f). \tag{3.2}$$

The sum on the right hand side of this equation is the sum in (3.1), so we now have to deal with the integral on the left hand side. We will deal with it piece by piece. The sections from H to A and from B to C cancel one another out, because they go in opposite directions, and also because $f(z) = f(z+1)$, as f is a modular function for $\mathrm{SL}_2(\mathbf{Z})$.

We now evaluate the integral across the top of $\mathcal{C}$, from A to B; by the chain rule, we see that

$$f'(z) = \frac{d}{dq} \tilde{f}(q) \frac{dq}{dz},$$

so we find by substituting $\tilde{f}(q)$ for $f(z)$ that the integral from A to B is equal to the following integral over the circle $\mathcal{D}$ of radius $e^{-2\pi T}$ with centre at 0:

$$\frac{1}{2\pi i} \int_A^B \frac{f'(z)}{f(z)} = \frac{1}{2\pi i} \int_{\mathcal{D}} \frac{d\tilde{f}/dq}{\tilde{f}(q)} dq;$$

as the integral goes round this circle in a clockwise direction (or, in the words of [Serre (1973a)], has a "negative orientation"), the integral evaluates to $-v_\infty(f)$.

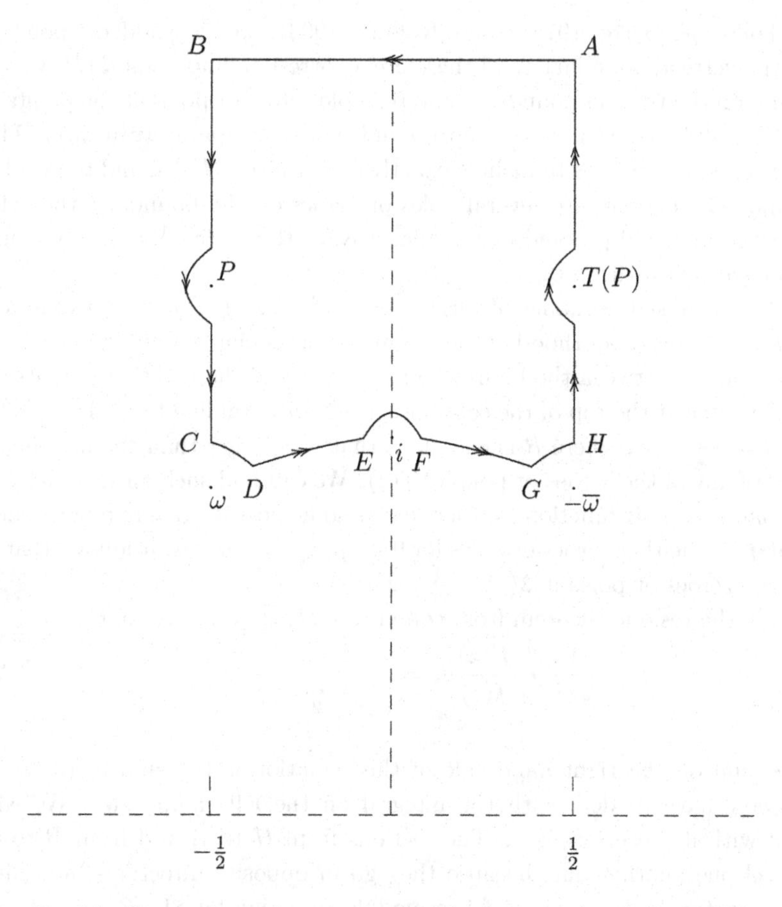

Fig. 3.1 Contour integral on the standard fundamental domain for $\mathrm{SL}_2(\mathbf{Z})$

We now consider the arcs from C to D and from G to H. From the derivation of the residue formula, we see that if $f(z)$ has Laurent expansion $a_n(z-h)^n+\cdots$ near the point $h\in\mathcal{H}$, with $a_n\neq 0$, then $f'(z)/f(z)$ has Laurent expansion $\frac{n}{z-h}+g(z)$, where $g(z)$ is a complex function which is holomorphic at h. If we now integrate $f'(z)/f(z)$ anticlockwise around a small circular arc of angle θ with centre h and radius ε (where "small" here means that there are no other zeroes or poles within the circle of which the arc is a part), then as we let $\varepsilon\to 0$ the integral tends to $mi\theta$.

We now apply this to the integral between C and D and let $\varepsilon\to 0$. The angle here tends towards $\pi/3$ and so we obtain

$$\frac{1}{2\pi i}\int_C^D \frac{f'(z)}{f(z)} \to -\frac{1}{2\pi i}\left(v_\omega(f)\cdot\frac{i\pi}{3}\right) = \frac{-v_\omega(f)}{6} \quad \text{as } \varepsilon \to 0.$$

By an exactly similar process we see that

$$\frac{1}{2\pi i}\int_G^H \frac{f'(z)}{f(z)} = \frac{-v_{-\bar{\omega}}(f)}{6} = \frac{-v_\omega(f)}{6};$$

the equality here follows because we can use the action of the matrix $\left(\begin{smallmatrix}0 & -1\\ 1 & 0\end{smallmatrix}\right)$ to relate the number of zeroes or poles of f at ω and at $-\bar{\omega}$.

Varying the method slightly, we find that the part of the integral from E to F (in other words, the small arc around i) evaluates to

$$\frac{1}{2\pi i}\int_E^F \frac{f'(z)}{f(z)} = -\frac{v_i(f)}{2}.$$

Finally, we must deal with the two arcs from D to E and from F to G. We will show that, as $\varepsilon \to 0$,

$$\frac{1}{2\pi i}\int_D^E \frac{f'(z)}{f(z)} + \frac{1}{2\pi i}\int_F^G \frac{f'(z)}{f(z)} \to \frac{k}{12}.$$

This will prove the theorem, as we will then have evaluated the entire integral. We will use the fact that the matrix S maps the arc DE onto the arc GF; from the transformation formula, we see that $f(Sz) = z^k f(z)$, and hence

$$\frac{f'(Sz)}{f(Sz)} = \frac{k}{z} + \frac{f'(z)}{f(z)}.$$

Using this, we see that

$$\begin{aligned}
\frac{1}{2\pi i}\int_D^E \frac{f'(z)}{f(z)}dz + \frac{1}{2\pi i}\int_F^G \frac{f'(z)}{f(z)}dz &= \frac{1}{2\pi i}\int_D^E \left(\frac{f'(z)}{f(z)} - \frac{f'(Sz)}{f(Sz)}\right) dz\\
&= -\frac{1}{2\pi i}\int_D^E k\frac{z'}{z}dz\\
&\to -k\left(-\frac{1}{12}\right) = \frac{k}{12}
\end{aligned}$$

as the radius of the arcs CD, EF and GH all tend to zero. The factor $1/12$ appears because in the limit we are integrating over the arc of the circle of radius 1 centred at the origin from ω to i, and this arc has length $\pi/6$. □

There are similar proofs for more general congruence subgroups, on the same principle. One constructs a fundamental domain F and then performs a similar contour integral along the boundary of F, using the same techniques to ensure that every pole or zero of the modular function f is included within the integral exactly once.

We note that although there are many choices of fundamental domain, for ease of computation we may wish to choose one that is connected, does not have closed sets removed, and so on. For instance, in our proof of the residue formula, we assumed that the fundamental domain was the "standard" one to make our computations easier.

The residue formula is very useful for direct computations; in the next section we will use it to prove formulae for the dimension of the spaces $M_k(\mathrm{SL}_2(\mathbf{Z}))$ and $S_k(\mathrm{SL}_2(\mathbf{Z}))$.

3.2 Explicit formulae for the dimensions of spaces of modular forms

In fact, we can do rather more than the abstract statement that the spaces of modular forms for $\mathrm{SL}_2(\mathbf{Z})$ are finite-dimensional. We can actually make these statements effective, by giving a finite set of generators for any space of modular forms of level 1, and proving explicit dimension formulae. We will then state similar dimension formulae for more general congruence subgroups.

3.2.1 *Formulae for the full modular group*

We will begin by proving some basic results about $M_k(\mathrm{SL}_2(\mathbf{Z}))$ and $S_k(\mathrm{SL}_2(\mathbf{Z}))$. These will allow us to prove dimension formulae and also to find a basis for any given space of modular forms for $\mathrm{SL}_2(\mathbf{Z})$.

Proposition 3.3. *Let k be an integer. Then*

(1) $M_0(\mathrm{SL}_2(\mathbf{Z})) = \mathbf{C}$.
(2) $M_2(\mathrm{SL}_2(\mathbf{Z})) = 0$, *and if $k < 0$ or if k is odd then* $M_k(\mathrm{SL}_2(\mathbf{Z})) = 0$.
(3) If $k \in \{4, 6, 8, 10, 14\}$, then $M_k(\mathrm{SL}_2(\mathbf{Z})) = \mathbf{C}E_k$.
(4) If $k < 12$ or $k = 14$ then $S_k(\mathrm{SL}_2(\mathbf{Z})) = 0$; $S_{12}(\mathrm{SL}_2(\mathbf{Z})) = \mathbf{C}\Delta$, *and if $k \geq 16$ then* $S_k(\mathrm{SL}_2(\mathbf{Z})) = \Delta \cdot M_{k-12}(\mathrm{SL}_2(\mathbf{Z}))$.
(5) If $k \geq 4$ then $M_k(\mathrm{SL}_2(\mathbf{Z})) = S_k(\mathrm{SL}_2(\mathbf{Z})) \oplus \mathbf{C}E_k$.

Proof.

(1) We saw above that the only possible modular forms of weight 0 are constants.
(2) There is no solution in non-negative integers to (3.1) when $k = 2$, or when $k < 0$. If k is odd then we have seen using the transformation formula in (2.5) that there are no nonzero forms of odd weight.
(3) If $k \in \{4, 6, 8, 10, 14\}$, then there is exactly one solution to (3.1), and we have already shown that E_k is a nonzero modular form of weight k. If $f \in M_k(\mathrm{SL}_2(\mathbf{Z}))$, then f/E_k is a modular form of weight 0, hence a constant, so clearly $f = c \cdot E_k$ for some $c \in \mathbf{C}$.
(4) If $k < 10$ or $k = 14$ then there are no solutions to (3.1) with $v_\infty(f) > 0$ (because the left-hand side is too large), so there can be no nonzero cusp forms of weight k. If $k = 12$ then there is exactly one solution to (3.1) in non-negative integers, which has $v_\infty(f) = 1$ and all other $v_P(f) = 0$, and we have already shown that Δ is a cusp form of weight 12, hence $S_{12}(\mathrm{SL}_2(\mathbf{Z})) = \mathbf{C}\Delta$.
Let $k = 12$ or $k \geq 16$. To prove the final assertion of this part, we note that $f \in S_k(\mathrm{SL}_2(\mathbf{Z}))$ must have at least a simple zero at ∞, by definition. We also note that the unique zero of the Δ function is at ∞, so f/Δ has no poles on $\mathcal{H}$ or at ∞, so it is therefore a modular form of weight $k - 12$, as was to be shown.
(5) Let $f \in M_k(\mathrm{SL}_2(\mathbf{Z}))$. If f is not a cusp form, then its Fourier expansion at ∞ has nonzero constant term c. We note that $f - cE_k$ has constant term 0, so vanishes at the cusp ∞, so is a cusp form.

□

We note here that we are genuinely using the fact that a modular form f is holomorphic both on $\mathcal{H}$ and at the cusp ∞; if we weaken these conditions then we will almost certainly have infinite-dimensional spaces. For example, consider the function $j(z) := E_4^3/\Delta$, which is a weight 0 weakly modular form, but is not a modular form because it has a simple pole at ∞. We will see later that j is an example of a *modular function*, and also that the space of modular functions of weight 0 is infinite-dimensional.

Remark 3.4. In the final part of this proposition, our result relies heavily on the fact that $\mathrm{SL}_2(\mathbf{Z})$ has only one cusp, which we can take to be ∞. For other congruence subgroups, it is extremely important to note that the vanishing of the constant term of the Fourier expansion at ∞ does *not* necessarily make a modular form a cusp form[2]. In Exercise 10 of Chapter 4, we

[2] In [Gouvêa and Mazur (1992)], the modular form $E_k(q) - E_k(q^p)$, for $k \geq 4$ and even, is called the *evil twin*, because it mimics a cusp form without being one in various ways,

will see an explicit example of a modular form (for a congruence subgroup) that vanishes at ∞ but is not a cusp form.

We now use Proposition 3.3 to derive a formula for the dimension of the space of modular forms of weight k for $\mathrm{SL}_2(\mathbf{Z})$.

Theorem 3.5 (Dimension of $M_k(\mathrm{SL}_2(\mathbf{Z}))$ and $S_k(\mathrm{SL}_2(\mathbf{Z}))$). *Let k be an even positive integer. Then*

$$\dim M_k(\mathrm{SL}_2(\mathbf{Z})) = \begin{cases} \left\lfloor \frac{k}{12} \right\rfloor + 1, & \text{if } k \not\equiv 2 \mod 12 \\ \left\lfloor \frac{k}{12} \right\rfloor, & \text{if } k \equiv 2 \mod 12. \end{cases} \tag{3.3}$$

and

$$\dim S_k(\mathrm{SL}_2(\mathbf{Z})) = \begin{cases} \left\lfloor \frac{k}{12} \right\rfloor, & \text{if } k \not\equiv 2 \mod 12 \\ \left\lfloor \frac{k}{12} \right\rfloor - 1, & \text{if } k \equiv 2 \mod 12. \end{cases} \tag{3.4}$$

Proof. We will proceed by induction. If $k \leq 14$, then we can use Proposition 3.3 to find the dimensions of $M_k(\mathrm{SL}_2(\mathbf{Z}))$ and $S_k(\mathrm{SL}_2(\mathbf{Z}))$ and verify that these are the numbers given by the formula. Also, by part 5 of Proposition 3.3, we see that the dimension of $S_k(\mathrm{SL}_2(\mathbf{Z}))$ is one less than the dimension of $M_k(\mathrm{SL}_2(\mathbf{Z}))$.

Now we assume that we have proved the theorem for weight k forms and we consider weight $k+12$ forms. From the preceding proposition we see that

$$M_{k+12}(\mathrm{SL}_2(\mathbf{Z})) = \mathbf{C}E_{k+12} \oplus S_{k+12}(\mathrm{SL}_2(\mathbf{Z})) = \mathbf{C}E_{k+12} \oplus \Delta \cdot M_k(\mathrm{SL}_2(\mathbf{Z})),$$

and now we are in a position to apply the induction hypothesis, because we know the dimension of $M_k(\mathrm{SL}_2(\mathbf{Z}))$. The cases $k \equiv 2 \mod 12$ and $k \not\equiv 2 \mod 12$ are exactly similar, and we see that $\lfloor k/12 \rfloor + 1 = \lfloor (k+12)/12 \rfloor$, so we are done. □

Using this theory, and the knowledge that E_4 and E_6 are modular forms, we can find a basis for $S_k(\mathrm{SL}_2(\mathbf{Z}))$ and $M_k(\mathrm{SL}_2(\mathbf{Z}))$.

Proposition 3.6. *Let $f(z) \in M_k(\mathrm{SL}_2(\mathbf{Z}))$ be a modular form. Then there exist $\alpha_{a,b} \in \mathbf{C}$ such that*

$$f(z) = \sum_{\substack{a,b \in \mathbf{N} \\ 4a+6b=k}} \alpha_{a,b} \cdot E_4(z)^a \cdot E_6(z)^b.$$

such as vanishing at the cusp ∞.

Proof. We will use induction on k. If $k \in \{0, 4, 6, 8, 10, 14\}$ then the space of modular forms is one-dimensional, and the following modular forms span the respective spaces $M_k(\mathrm{SL}_2(\mathbf{Z}))$: 1, E_4, E_6, E_4^2, $E_4 \cdot E_6$, $E_4^2 \cdot E_6$. We see that all of these are polynomials in E_4 and E_6. Also, if $k = 12$ then a basis is given by $\{E_4^3, E_6^2\}$.

Now we assume that $k \geq 16$, and that if $k' < k$ then every element of $M_{k'}(\mathrm{SL}_2(\mathbf{Z}))$ can be written as a polynomial in E_4 and E_6. It is clearly possible to find integers a and b such that $4a + 6b = k$, so we have a non-cuspidal modular form $E_4^a \cdot E_6^b$ of weight k with constant term 1. Let f be an arbitrary nonzero modular form of weight k. We can find $c \in \mathbf{C}$ such that $f - cE_4^a \cdot E_6^b \in S_k(\mathrm{SL}_2(\mathbf{Z}))$ (we take c to be the constant coefficient in the Fourier expansion of f). We can write $f - cE_4^a \cdot E_6^b = \Delta \cdot g$, where Δ is the normalized cusp form of weight 12 and level 1, and g is a modular form of weight $k - 12$. Using the induction hypothesis, we see that g is a polynomial in E_4 and E_6, and as $\Delta = (E_4^3 - E_6^2)/1728$, it is clear that f is a polynomial in E_4 and E_6. □

Remark 3.7. An induction of this form is a common method of proof for results of the form "Property X holds for all $M_k(\mathrm{SL}_2(\mathbf{Z}))$". Exercise 1 gives an example of another, similar, proof. A common feature is the use of induction, and the bijection from $M_k(\mathrm{SL}_2(\mathbf{Z}))$ to $S_{k+12}(\mathrm{SL}_2(\mathbf{Z}))$ given by multiplication by Δ.

3.2.2 *Formulae for congruence subgroups*

Having given a full account of the induction for $\mathrm{SL}_2(\mathbf{Z})$, we will not give all the details for general congruence subgroups, but will state dimension formulae for congruence subgroups and give an overview of the proof. These formulae can be found in several places; Sections 3.5 and 3.6 of [Diamond and Shurman (2005)], Sections 6.1 and 6.2 of [Stein (2007)] and Section 2.6 of [Shimura (1994)], for instance. An older reference is the paper [Cohen and Oesterlé (1977)].

First we will need to define some notation. If p is a prime and N is a positive integer, we define $v_p(N)$ to be the normalized p-valuation of N.

We define the following functions:

$$\mu_{0,2}(N) := \begin{cases} 0, & \text{if } 4|N, \\ \prod_{\substack{p|N \\ p \text{ prime}}} \left(1+\left(\frac{-4}{p}\right)\right) & \text{otherwise.} \end{cases}$$

$$\mu_{0,3}(N) := \begin{cases} 0 & \text{if } 2|N \text{ or } 9|N, \\ \prod_{\substack{p|N \\ p \text{ prime}}} \left(1+\left(\frac{-3}{p}\right)\right) & \text{otherwise.} \end{cases}$$

$$c(N) := \sum_{d|N} \phi(\gcd(d, N/d)).$$

Here ϕ is the standard Euler ϕ function, also called the Euler totient function, $c(N)$ is the number of cusps of $\Gamma_0(N)$, and $\left(\frac{\cdot}{p}\right)$ is the Legendre symbol if $p \neq 2$ and is -1 if $p = 2$. We define $g(N)$ to be

$$g(N) := 1 + \frac{[\mathrm{SL}_2(\mathbf{Z}) : \Gamma_0(N)]}{12} - \frac{\mu_{0,2}(N)}{4} - \frac{\mu_{0,3}(N)}{4} - \frac{c(N)}{2}.$$

($g(N)$ is known as the *genus of the modular curve* $X_0(N)$; this is an object that parametrizes pairs of elliptic curves together with a cyclic subgroup of order N. We briefly mention these objects in Remark 5.15, but will not go into the theory of modular curves here; a good first reference is [Diamond and Im (1995)], Section 7).

We can now give the general dimension formula for spaces of modular forms for $\Gamma_0(N)$.

Theorem 3.8. *Let k be a non-negative even integer and let $\Gamma_0(N)$ be a congruence subgroup. If $k = 0$, we have $\dim S_k(\Gamma_0(N)) = 0$, if $k = 2$, then we have $\dim S_k(\Gamma_0(N)) = g(N)$, and if $k \geq 4$ then*

$$\dim S_k(\Gamma_0(N)) = (k-1)(g(N)-1) + \left(\frac{k}{2} - 1\right) \cdot c(N) + \mu_{0,2}(N) \cdot \left\lfloor \frac{k}{4} \right\rfloor + \mu_{0,3} \cdot \left\lfloor \frac{k}{3} \right\rfloor.$$

The dimension of the space of Eisenstein series (the Eisenstein subspace*) is*

$$\dim E_k(\Gamma_0(N)) = \begin{cases} c(N) & \text{if } k \neq 2, \\ c(N) - 1 & \text{if } k = 2. \end{cases}$$

Remark 3.9. We see that in the special case where $N = 1$, we have that $g(1) = 0$, and the formula in Theorem 3.8 simplifies to $\lfloor k/12 \rfloor$ if $k \not\equiv 2 \mod 12$ and $\lfloor k/12 \rfloor - 1$ if $k \equiv 2 \mod 12$.

There are similar formulae for the dimension of the space of modular forms for $\Gamma_1(N)$; we have to define some more functions.

$$\mu_1(N) := \begin{cases} [\mathrm{SL}_2(\mathbf{Z}) : \Gamma_0(N)] & \text{if } N = 1,2, \\ [\mathrm{SL}_2(\mathbf{Z})/\{\pm 1\} : \Gamma_1(N)/\{\pm 1\}] & \text{otherwise.} \end{cases}$$

$$\mu_{1,2} := \begin{cases} 0 & \text{if } N \geq 4, \\ \mu_{0,2}(N) & \text{otherwise.} \end{cases}$$

$$\mu_{1,3}(N) := \begin{cases} 0 & \text{if } N \geq 4, \\ \mu_{0,3}(N) & \text{otherwise.} \end{cases}$$

$$C(N) := \begin{cases} c(N) & \text{if } N = 1,2, \\ 3 & \text{if } N = 4, \\ \sum_{d|N} \frac{\phi(d)\phi(N/d)}{2} & \text{otherwise.} \end{cases}$$

We also define

$$G(N) = 1 + \frac{\mu_1(N)}{12} - \frac{\mu_{1,2}(N)}{4} - \frac{\mu_{1,3}}{3} - \frac{C(N)}{2}.$$

(This is the genus of the modular curve $X_1(N)$, which parametrizes pairs of elliptic curves together with points of order N).

We can now give the dimension formulae for modular forms for $\Gamma_1(N)$.

Theorem 3.10. *Let N and k be positive integers, and let $k \geq 2$. If $N = 1,2$, then $\dim S_k(\Gamma_1(N)) = \dim S_k(\Gamma_0(N))$, because $\Gamma_1(N) = \Gamma_0(N)$. If $N \geq 3$ then the dimension is given by*

$$\dim S_k(\Gamma_1(N)) = \begin{cases} G(N) & \text{if } k = 2, \\ (k-1)(G(N)-1) + \left(\frac{k}{2} - 1\right) C(N) + \alpha, & \text{if } k \geq 3, \end{cases}$$

where

$$\alpha = \begin{cases} \frac{1}{2} & \text{if } N = 4 \text{ and } k \text{ is odd,} \\ \left\lfloor \frac{k}{3} \right\rfloor & \text{if } N = 3, \\ 0 & \text{otherwise.} \end{cases}$$

The dimension of the space of Eisenstein series is

$$\dim E_k(\Gamma_1(N)) = \begin{cases} C(N) & \text{if } k \geq 3, \\ C(N) - 1 & \text{if } k = 2. \end{cases}$$

Remark 3.11. We notice that in the statement of the theorem there are some explicit identities of the form $S_k(\Gamma_0(N)) = S_k(\Gamma_1(N))$ given; there are a handful of others. It can be seen that if k is even, then $S_k(\Gamma_1(4)) = S_k(\Gamma_0(4))$; we can show this either by checking that their dimensions are equal (and noting that one is contained within the other), or by noting that there are only two Dirichlet characters modulo 4; one is odd and one is even, and the even one is trivial; see Exercise 21.

Remark 3.12. There is an extra term α in the formula when $N \leq 4$; this appears because there are two types of cusps; *regular* and *irregular*, and these contribute different amounts to the dimension sum. We will not discuss these here; references can be found in [Diamond and Shurman (2005)], Section 3.2, and [Koblitz (1993)], Exercise III.3.2. An example of an irregular cusp is the cusp $-1/2$ for the congruence subgroup $\Gamma_1(4)$.

We notice also that the dimensions of the spaces of Eisenstein series are (for $k \neq 2$) equal to the number of cusps.

The basic idea of the proofs for these formulae is the same as for the case of $\mathrm{SL}_2(\mathbf{Z})$; we can use results we have already derived to find fundamental domains for our congruence subgroups, and then we can perform contour integrals around these fundamental domains to obtain residue formulae akin to that given as (3.1). We can derive formulae from this.

Proofs of these results about the dimensions of spaces of modular forms can be found in Sections 3.5 and 3.6 of [Diamond and Shurman (2005)]; the specific question of the dimension of the Eisenstein subspace is also discussed in [Koblitz (1993)], at the end of Chapter III, Section 5.

Analogous formulae can be generated for other congruence subgroups, such as $\Gamma(N)$ or $\Gamma^0(N)$, in a similar way.

3.3 The Sturm bound

If we are given two modular forms of the same weight, how can we tell if they are the same? We can consider the Fourier expansion of their difference and see whether the first few coefficients are zero, but we need to use some thought to see whether this means that the modular form itself is zero. For instance, if we look at the modular form Δ^{100} we see that its first 100 Fourier coefficients vanish, but this does not mean that it is zero.

In this section, we will derive an explicit and effective bound on the number of coefficients of a modular form that one has to compute to show that it is zero. We note that we do need to know both the weight *and* the level of the modular form to do this, as otherwise we will never know if we have computed enough terms; see Exercise 4 for more details.

We can use the residue formula (3.1) to settle this directly for modular forms for $\mathrm{SL}_2(\mathbf{Z})$. If the order of vanishing $\nu_\infty(f)$ of a modular form f is strictly greater than $k/12$ then there is no possible solution to the residue formula equation, so f must be identically zero.

The general theorem was first given by Sturm in [Sturm (1987)]; it is called either the *Sturm bound* or the *Hecke bound* in the literature. A more modern reference is Section 9.4 of [Stein (2007)].

Theorem 3.13 (Sturm [Sturm (1987)]). *Let $\Gamma \in \mathrm{SL}_2(\mathbf{Z})$ be a congruence subgroup of index M and let $f \in M_k(\Gamma)$ be a modular form. If*

$$\nu_\infty(f) > M \cdot \frac{k}{12}$$

then f is identically zero.

Proof. If $\Gamma = \mathrm{SL}_2(\mathbf{Z})$ then we can prove this immediately by considering the residue theorem (3.1), as we did above; if f has a zero of order greater than $k/12$ at ∞, then it must be zero, because the right-hand side of the residue formula (3.1) is $k/12$, and all of the elements of the left-hand side are non-negative.

We now use the fact that Γ is a subgroup of finite index of $\mathrm{SL}_2(\mathbf{Z})$; this means that we can write

$$\mathrm{SL}_2(\mathbf{Z}) = \bigcup_{i=1}^{M} \Gamma\gamma_i,$$

for some finite set of $\gamma_i \in \mathrm{SL}_2(\mathbf{Z})$. Without loss of generality, we can assume that $\gamma_1 = I$. We define a modular form F by

$$F := f \cdot \prod_{i=2}^{M} f|[\gamma_i]_k;$$

we will now show that F is a modular form for $\mathrm{SL}_2(\mathbf{Z})$, so we reduce it to a case that we have solved. We note that each of the $f|[\gamma_i]$ is a modular form of weight k for a suitable congruence subgroup, by a variant of the argument before Proposition 2.20.

To show that F is a modular form for $\mathrm{SL}_2(\mathbf{Z})$, we need to show that it transforms correctly under the action of $\mathrm{SL}_2(\mathbf{Z})$. This will follow because allowing an element $g \in \mathrm{SL}_2(\mathbf{Z})$ to act on the right permutes the γ_i, so the product is left unchanged. This means that F is not just a modular form for Γ, but a modular form for the full modular group.

We see that F has weight kM, because it is the product of M weight k modular forms, so we now apply the theorem in the level 1 case to obtain the bound. □

Remark 3.14. We see that the bound is in fact sharp in general for $\mathrm{SL}_2(\mathbf{Z})$; if we consider the modular form Δ^i, we see that it has a unique zero of exact multiplicity i at ∞, so we cannot replace the strict inequality with a non-strict inequality.

We can now prove as a corollary and special case of this result that, for every congruence subgroup Γ, the space of modular forms of weight 0 is exactly the constant functions. We will need this specific result in Chapter 5.

Corollary 3.15. *Let Γ be a congruence subgroup, and let $f \in M_0(\Gamma)$. Then f is a constant function.*

Proof. If $\Gamma = \mathrm{SL}_2(\mathbf{Z})$ then we have proved this result above as Proposition 3.3, so we will assume that Γ is strictly contained in $\mathrm{SL}_2(\mathbf{Z})$.

We let $f \in M_0(\Gamma)$, and let $a = f(Z)$ be the value of f at some fixed point $Z \in \mathcal{H}$. Because Γ is a congruence subgroup, it has finite index within the full modular group $\mathrm{SL}_2(\mathbf{Z})$, so we can write $\mathrm{SL}_2(\mathbf{Z})$ as a finite union of cosets of the form $\alpha_i\Gamma$. We now consider the modular function of weight 0

$$F(z) := \prod_{i=1}^{n}(f|[\alpha_i^{-1}]_0 - a) = \prod_{i=1}^{n}\left(f(\alpha_i^{-1}z) - a\right).$$

By the arguments given in Theorem 3.13, we see that F is a modular form of weight 0 for $\mathrm{SL}_2(\mathbf{Z})$, so by Proposition 3.3, we see that F is a constant.

We now note that we can choose one of the cosets to be $I\Gamma$, where I is the identity matrix. The term in the product defining F which corresponds to I will simply be $(f(z) - a)$, as the identity matrix acts trivially. We now compute $F(Z)$ for our chosen $Z \in \mathcal{H}$, and we see that $F(Z) = 0$, so in fact F is identically zero. This means that one of the factors of F must be zero, as the meromorphic functions on $\mathcal{H}$ are a field (Exercise 13).

In other words, we have that $f(\alpha_i^{-1}z) - a = 0$ for some i and for all z in the Poincaré upper half plane. By making the change of variable from z to $\alpha_i z$ (which is allowable, because if $z \in \mathcal{H}$ then $\alpha_i z \in \mathcal{H}$ as well), we see that $f(z) = a$ for all $z \in \mathcal{H}$, so f is a constant function, which is what we wanted to prove. □

There is an interesting distinction here between the worst-case scenario, which is that the Sturm bound is sharp, and the fact that modular forms, and especially eigenforms for the Hecke operators (which we will see in Chapter 4), found "in nature" are often very easy to tell apart; often, any difference between the forms appears well before the Sturm bound[3].

Using the methods of this section, we can see that there is a similar bound on the number of zeroes that a modular form can have at any particular point of $\mathcal{H}$. We have presented our work in this way because we can

[3]This type of analysis often occurs in computer science, where there are several algorithms whose worst-case behaviour is markedly worse than the usual behaviour; examples are the "quicksort" algorithm for sorting a list and the simplex algorithm to solve linear programming problems.

find a Fourier expansion at ∞ for our modular forms, so we already know what the order of vanishing of f at ∞ is.

The existence of the Sturm bound for arbitrary congruence subgroups actually implies the finite-dimensionality theorem we stated earlier; we now give the proof.

Proof. [Proof of Theorem 3.1] Let Γ be a congruence subgroup of index M, and let $f \in M_k(\Gamma)$ be a nonzero modular form. If $k = 0$, we see that f must be a constant by applying Theorem 3.13, and if $k < 0$ then f is identically zero; this follows because the modular form F we would obtain by following the proof of the Sturm bound would have weight $mk < 0$, and would therefore be identically zero, and therefore at least one of the $f|[\gamma_i]_k$ would be zero. We can therefore assume that f must have weight greater than 0.

This means that there are only finitely many choices for the order of vanishing of f at ∞, and if we have two modular forms which have the same order of vanishing r at ∞ then we can take a suitable linear combination of them, which will have order of vanishing strictly greater than r; we can see this by a comparison of their Fourier series at ∞. This means that we can write down a basis for the space with dimension at most the number of possible orders of vanishing, and this is finite. □

We see that we have actually proved more than the bare statement of Theorem 3.1, as we have the explicit bound of $kM/12$ on the dimension. This is compatible with the exact dimension formulae we stated; our proof of Theorem 3.1 does not show that the bound of $kM/12$ is obtained.

3.4 Exercises

(1) Using induction, or otherwise, show that every modular form of level 1 with integer coefficients is a polynomial in E_4, E_6 and Δ *with integral coefficients* (where we recall that these are normalized as follows: $E_4(q) = 1 + \cdots$, $E_6(q) = 1 + \cdots$, and $\Delta = q + \cdots$).

(2) In this exercise we will show that all modular forms of level $\Gamma_0(2)$ can be written as polynomials in two modular forms, $E_2^*(q) := E_2(q) - 2E_2(q^2)$ and $E_4(q)$.

   (a) Check that $M_2(\Gamma_0(2))$ has dimension 2.
   (b) Check that $(E_2^*)^2$ and E_4 are linearly independent modular forms of weight 4. Check also that $M_6(\Gamma_0(2))$ has dimension 2, and find

a basis in terms of E_2^* and E_4.

(c) Show that $S_8(\Gamma_0(2))$ has dimension 1, and use Theorem 5.10 to find a nonzero element F in this space. Write F in terms of E_2^* and E_4.

(d) Show that $S_k(\Gamma_0(2)) = F \cdot M_{k-8}(\Gamma_0(2))$.

(e) Show that the "Eisenstein" subspace of $M_k(\Gamma_0(2))$ has dimension 2; in other words, that we have

$$M_k(\Gamma_0(2)) = \mathbf{C}A \oplus \mathbf{C}B \oplus S_k(\Gamma_0(2)),$$

where A and B are linearly independent non-cuspidal modular forms of weight k. You may wish to consider how many cusps $\Gamma_0(2)$ has; does this knowledge help you here?

(f) Hence or otherwise, show that the following formulae hold:

$$\dim M_k(\Gamma_0(2)) = \left\lfloor \frac{k}{4} \right\rfloor + 1,$$

$$\dim S_k(\Gamma_0(2)) = \left\lfloor \frac{k}{4} \right\rfloor - 1.$$

(g) Check that these formulae are consistent with those given in Theorem 3.8.

(3) Choose a congruence subgroup Γ and perform similar computations to those given in Exercise 2.

(4) Find modular forms of fixed weight which have arbitrarily many zeroes at the cusp ∞. Find modular forms of fixed level which have arbitrarily many zeroes at ∞.

Chapter 4

The arithmetic of modular forms

> There still remain three studies suitable for free citizens. Arithmetic is one of them. — Plato, the Laws.

From our discussions in Chapter 2 and Chapter 3, one could be forgiven for thinking that, in the words of [Stein (2007)], modular forms live in "an obscure corner of complex analysis". For number theorists, however, modular forms are interesting for their arithmetic properties. In Chapter 2, we saw hints of this, with the constant term of $G_k(z)$ being shown to be $2\zeta(k)$, the appearance of the Bernoulli numbers with their link to Fermat's Last Theorem, and the terms of the Fourier expansion of $E_k(z)$ being multiplicative functions.

We will study these properties in more detail in this chapter, in preparation for the applications of the theory to arithmetic questions which we will see in Chapter 5. Much of what we will see and prove will depend on the finite-dimensionality results we have proved in Chapter 3.

In addition to the books mentioned at the beginning of Chapter 2, another and more advanced reference for the arithmetic aspects of modular forms is [Shimura (1994)]. The introductory chapter of [Bump (1997)] is also useful, as are the books [Sarnak (1990)] and [Hellegouarch (2002)].

The book [Miyake (2006)] considers the subject in much more generality; the congruence subgroups are examples of *Fuchsian groups*, and one can develop the theory for these groups. The book [Ford (1957)] gives a classical treatment of the subject of automorphic forms in this generality.

The online notes of Milne on modular forms [Milne (1997b)] and elliptic curves (available as a book [Milne (2006)] and online as [Milne (1997a)] are also useful and freely available references for the arithmetic aspects of the theory.

It is a recurrent theme of books on modular forms that no one book covers everything, but by reading several books one can obtain an overview of the subject; the different (possible) approaches between them cover nearly every way in which we can approach modular forms.

4.1 Hecke operators

There are a collection of extremely important and fundamental linear operators acting on modular forms, called the *Hecke operators.* In this section, we will motivate them, define them and talk about their properties.

4.1.1 *Motivation for the Hecke operators*

If we take coprime positive integers m and n then we can check experimentally that $\tau(m \cdot n) = \tau(m) \cdot \tau(n)$ (and in fact this holds in general); we say that such a function is *multiplicative.* Also, it is a pleasant (and much more elementary) exercise (Exercise 3) to show that the coefficients of the Eisenstein series E_k are multiplicative in this sense. It is natural to ask how one can generalize this, and maybe even to ask if one can find a basis of the space of modular forms composed of such forms.

It was noticed long ago (by Ramanujan, amongst others) that the Fourier coefficients of Δ and of the E_k also satisfy recurrence relations; for instance, if p is a prime, then we have

$$\tau(p^{r+2}) = \tau(p^{r+1}) \cdot \tau(p) - p^{11}\tau(p^r), \text{ for } r \geq 0, \tag{4.1}$$

and if $k \geq 4$ is an even integer, we have

$$\sigma_{k-1}(p^{r+2}) = \sigma_{k-1}(p^{r+1}) \cdot \sigma_{k-1}(p) - p^{k-1}\sigma_{k-1}(p^r), \text{ for } r \geq 0.$$

In [Hellegouarch (2002)], Section 5.7, the following ingenious heuristic argument is given for the $\tau(n)$ to be multiplicative. Let r be a positive integer, and define the operator U_r on Fourier expansions by

$$U_r\left(\sum_{n=0}^{\infty} a_n q^n\right) := \sum_{n=0}^{\infty} a_{rn} q^n.$$

If U_r was an endomorphism of $S_{12}(\mathrm{SL}_2(\mathbf{Z}))$ (which, sadly, it is not) then we would have that

$$U_r\Delta = c_r\Delta \text{ for some } c_r \in \mathbf{C},$$

because $S_{12}(\mathrm{SL}_2(\mathbf{Z}))$ is a 1-dimensional complex vector space. Furthermore, by looking at the explicit action of U_r, we see that we would have $\tau(r) =$

$c_r\tau(1) = c_r$ for every $r \geq 1$, and furthermore, by iterating this, we would find that

$$\tau(mn) = \tau(m) \cdot c_r = \tau(m)\tau(n),$$

for every $m, n \in \mathbf{N}$ (so we would actually have that τ is completely multiplicative, not just multiplicative).

We will modify this idea of using U_r to show multiplicativity, and introduce endomorphisms of the spaces of modular forms $M_k(\mathrm{SL}_2(\mathbf{Z}))$ and $S_k(\mathrm{SL}_2(\mathbf{Z}))$ which are "averages" in some sense of the U_r; these will allow us to find modular forms which have Fourier coefficients with the multiplicative property. We will also see that the operator U_r is a map from certain spaces of modular forms to themselves, but not normally modular forms for $\mathrm{SL}_2(\mathbf{Z})$.

It is interesting to think about how one would find out that certain modular forms satisfy recurrence relations such as (4.1), if one had to reinvent the theory from scratch. When Ramanujan made his conjectures there was very little computing power available, so making large tables of coefficients was rather harder than it would be today. Noticing that the $\tau(n)$ are multiplicative requires much less data. Both of these could be guessed at by analogy with the coefficients of the Eisenstein series E_k, although the proof that the Fourier coefficients of E_k are multiplicative and satisfy the recurrence relation uses only elementary methods.

4.1.2 *Hecke operators for* $M_k(\mathrm{SL}_2(\mathbf{Z}))$

We first consider the simplest case; Hecke operators acting on modular forms for the full modular group $\mathrm{SL}_2(\mathbf{Z})$ (also known in the literature as the "level 1" case). There are several equivalent ways to consider this; we follow the approach of [Hellegouarch (2002)], Section 5.7.1, which uses coset operators defined in terms of matrices of determinant p to define the Hecke operator T_p. In a later section, we will use a definition in terms of double cosets; see [Koblitz (1993)], Section III.5 for yet another approach, involving modular points. We stress that these different approaches do all give the same Hecke operators.

Definition 4.1 (Hecke operators). *Let n be a positive integer and let M be the set of matrices $\left(\begin{smallmatrix} a & b \\ c & d \end{smallmatrix}\right) \in M_2(\mathbf{Z})$ with determinant n. We define the* Hecke operator T_n *to be*

$$(T_n f)(z) := n^{k-1} \sum_{\mu \in \mathrm{SL}_2(\mathbf{Z}) \backslash M} f|[\mu]_k. \tag{4.2}$$

We have several things to check here; $\mathrm{SL}_2(\mathbf{Z})$ acts on M on the left, so we can decompose M into orbits under this action. There are finitely many of these orbits, so for any choice of the μ that we make, the sum will be finite and therefore will make sense.

Now we need to show that this operator is well-defined; in other words, the sum in (4.2) should be independent of the choice of orbits that we make. Let $\gamma \in \mathrm{SL}_2(\mathbf{Z})$. Then

$$f|[\gamma\mu]_k = (f|[\gamma]_k)|[\mu]_k = f|[\mu]_k,$$

because, as f is a modular form of weight k, $f|[\gamma]_k = f$. This means that it is well-defined, which is what we wanted to show.

We now have enough information to prove the following theorem, which amongst other things tells us that the T_n are multiplicative; this will allow us to prove that the Fourier coefficients of Δ are multiplicative.

Theorem 4.2. *Let k and n be positive integers, and let $f \in M_k(\mathrm{SL}_2(\mathbf{Z}))$ have Fourier expansion at ∞ given by $f(z) = \sum_{n=0}^{\infty} a_n q^n$. Then the following hold:*

(1) $T_n f \in M_k(\mathrm{SL}_2(\mathbf{Z}))$, and if $f \in S_k(\mathrm{SL}_2(\mathbf{Z}))$, then $T_n f \in S_k(\mathrm{SL}_2(\mathbf{Z}))$.
(2) If m and n are positive integers, then

$$T_n \cdot T_m = \sum_{0<d|(m,n)} d^{k-1} \cdot T_{nm/d^2} = T_m \cdot T_n. \tag{4.3}$$

In particular, we see that $T_m \cdot T_n = T_{mn}$ if $(m,n) = 1$.

Proof. First, we will prove that $T_n f$ is a modular form, by checking the three conditions given in the definition of a modular form in Chapter 2 (holomorphic on $\mathcal{H}$ and at ∞, and transforming correctly under the action of $\mathrm{SL}_2(\mathbf{Z})$).

We see from Definition 4.1 that if f is holomorphic on $\mathcal{H}$, then $T_n f$ is also holomorphic on $\mathcal{H}$; this is one of the three conditions that we have to check. We also see that $T_n f$ is weakly modular of weight k for $\mathrm{SL}_2(\mathbf{Z})$, so now we have to show that it is holomorphic at ∞.

We can choose the coset representatives μ to be upper triangular by multiplying by a suitable element of $\mathrm{SL}_2(\mathbf{Z})$ on the left, which will not change the sum in $T_n f$. We then multiply on the left by $\pm\left(\begin{smallmatrix} 1 & r \\ 0 & 1 \end{smallmatrix}\right)$, where $r \in \mathbf{Z}$. Because $ad = n$ (the matrix has determinant n) we can assume that $a > 0$ and $0 \le b < d$. Therefore, with this choice of representatives, we have that

$$(T_n f)(z) = n^{k-1} \sum_{\substack{a,d>0 \\ ad=n}} \sum_{b=0}^{d-1} d^{-k} f\left(\frac{az+b}{d}\right).$$

We will now use the following relation:

$$D := \sum_{b=0}^{d-1} f\left(\frac{az+b}{d}\right) = \sum_{m=0}^{\infty} d \cdot a_{md} \cdot q^{ma}.$$

This follows from the fact that the map $z \mapsto (az+b)/d$ sends q^r to $q^{ar/d} \cdot e^{2\pi irb/d}$; we collect up the powers of $e^{2\pi i}$ and note that if $d \nmid r$ then the coefficient of q^r in D will be 0, because the sum over the roots of unity will be 0. If on the other hand we have that $d \mid n$, the roots of unity are all going to be 1, so then we get d copies of a_r.

We see that if f is holomorphic at ∞, then $T_n f$ is also holomorphic at ∞. The constant term of D is given by $d \cdot a_0$, so f vanishes at ∞ if and only if $T_n f$ does, so the Hecke operators T_n preserve $S_k(\mathrm{SL}_2(\mathbf{Z}))$.

We now prove the second part of the theorem about how these Hecke operators combine with one another. Firstly, we will show that the special case where $(m,n) = 1$ holds; in other words, that $T_{mn} = T_m \cdot T_n$ when $(m,n) = 1$. We will then show (4.3) holds if m and n are both powers of the same prime p, and then build up the general case from these two special cases. We give an overview of this, and leave the details to Exercise 4.

The case when m and n are coprime follows because any prime power divisor of mn is either a divisor of m or a divisor of n. We can check that given a matrix from T_{mn}, we can write it as a product of a matrix representing T_m and a matrix representing T_n, and also that the Möbius transformations combine correctly. We see from Exercise 4 that each of the terms in the expansion of T_{mn} can be rewritten as a product of a term from the expansion of T_m and a term from the expansion of T_n, and this allows us to rewrite T_{mn} as the product of T_m and T_n.

We now show that

$$T_{p^r} \cdot T_{p^s} = \sum_{0<p^i \mid (p^r,p^s)} p^{i(k-1)} \cdot T_{p^{r+s-i}}, \tag{4.4}$$

if r and s are positive integers. We will do this by induction on s. Firstly, we see that $T_{p^r} \cdot T_p = T_{p^{r+1}} + p^{k-1} T_{p^{r-1}}$ by multiplying out the products on both sides, and checking that they are equal. We then assume (4.3) for T_{p^r} and T_{p^s}, and use this to show that it holds for T_{p^r} and $T_{p^{s+1}}$.

Finally, we see that given general m and n, we can write T_m and T_n as products of Hecke operators $T_{p_i^{r_i}}$. Because these are coprime to one another if $p_i \neq p_j$, we can reorder them so that we have

$$T_{mn} = \prod_{i=1}^{t} T_{p_i^{r_i}},$$

where all of the p_i are distinct. We then use the fact that we have proved the result for T_{p^r}, substitute in the formula in (4.4), and we have the formula (4.3), as required. □

We note that if we take $m = p$ and $n = p^r$ in (4.3) and rearrange the terms, we obtain

$$T_{p^{r+1}} = T_{p^r} \cdot T_p - p^{k-1} T_{p^{r-1}}, \text{ for } r \geq 1. \tag{4.5}$$

This is the standard way that we build up Hecke operators; if we know how T_p acts for every p, then we can use (4.5) to compute all of the T_{p^r} and then every Hecke operator T_s by using that $T_{mn} = T_m \cdot T_n$. A similar form of this holds in more generality, which we consider in the next section.

We note that, because the Hecke operators map the space $M_k(\mathrm{SL}_2(\mathbf{Z}))$ to itself, and they preserve $S_k(\mathrm{SL}_2(\mathbf{Z}))$, if we have a 1-dimensional space of modular forms then its generator must be an eigenfunction for all of the Hecke operators, which we will call an *eigenform.* Let us now make a formal definition of this concept.

Definition 4.3. Let f be a modular form. If for every positive integer n there exists $\lambda_n \in \mathbf{C}$ such that $T_n f = \lambda_n f$, then we say that f is an *eigenform (for the Hecke operators T_n)*; we also sometimes say that it is a *simultaneous eigenform* for the T_n.

The canonical first example of a cuspidal eigenform is the Δ function; because $S_k(\mathrm{SL}_2(\mathbf{Z})) = \mathbf{C} \cdot \Delta$, we have that

$$T_n \Delta = \lambda_n \cdot \Delta, \text{ for all } n \in \mathbf{N}.$$

We see that this is the correct generalization to the motivational statements in the introduction involving U_r. Similar statements apply to all of the 1-dimensional spaces of cuspforms for $\mathrm{SL}_2(\mathbf{Z})$, by exactly similar reasoning.

The multiplicativity for the Fourier coefficients of the Eisenstein series E_k is a less deep fact; we can show it by explicitly considering the function $\sigma_{k-1}(n)$ and showing that it is multiplicative on prime powers.

4.1.3 *Hecke operators for congruence subgroups*

As we remarked above, there are several ways to consider the Hecke operators; another fruitful way of looking at them is to consider the theory of double cosets. We follow the treatment given in [Diamond and Shurman (2005)], Chapter 5 and [Koblitz (1993)], Section III.5 here. Let Γ_1 and Γ_2 be congruence subgroups of $\mathrm{SL}_2(\mathbf{Z})$ ([Koblitz (1993)] treats this in slightly more generality, which we will not need here).

Let α be an element of $\mathrm{GL}_2^+(\mathbf{Q})$ (this is the subgroup of elements of $\mathrm{GL}_2(\mathbf{Q})$ which have positive determinant). We define a *double coset in* $\mathrm{GL}_2^+(\mathbf{Q})$ to be

$$\Gamma_1\alpha\Gamma_2 = \{g_1\alpha g_2 : \, g_1 \in \Gamma_1, \, g_2 \in \Gamma_2\}.$$

We can use these double cosets to map modular forms to modular forms in the following way; we define the *double coset operator* by:

$$f|[\Gamma_1\alpha\Gamma_2]_k := \sum_i f|[\beta_j]_k,$$

where the β_j are a set of orbit representatives for $\Gamma_1\backslash\Gamma_1\alpha\Gamma_2$; in other words, we have $\Gamma_1\alpha\Gamma_2 = \cup_j \Gamma_1\beta_j$. We will now show that this double coset operator is well-defined, so that it is independent of the choice of the β_j, and that this action sends modular forms for Γ_1 to modular forms for Γ_2.

We need to show that if $f \in M_k(\Gamma_1)$, then $f|[\Gamma_1\alpha\Gamma_2]_k \in M_k(\Gamma_2)$, and that if f is a cusp form then $f|[\Gamma_1\alpha\Gamma_2]_k$ is a cusp form as well. This is given on pages 165–166 of [Diamond and Shurman (2005)]; we will summarize their argument here.

Firstly, we show that this action is well-defined. Suppose that we have two representatives of the same orbit in $\Gamma_1\backslash\Gamma_1\alpha\Gamma_2$, so $\Gamma_1\beta_1 = \Gamma_1\beta_2$. We can write these β_i as $\gamma_{i,1}\alpha\gamma_{i,2}$ for some $\gamma_{i,j} \in \Gamma_j$, and we see that $\alpha\gamma_{1,2} \in \Gamma_1\alpha\gamma_{2,2}$. Because f is a modular form for Γ_1, it is weight-k invariant under the action of elements of Γ_1, so we have that

$$f|[\beta_1]_k = f|[\alpha\gamma_{1,2}]_k = f|[\alpha\gamma_{2,2}]_k = f|[\gamma_{2,1}\alpha\gamma_{2,2}]_k = f|[\beta_2]_k,$$

as required. We may insert $\gamma_{2,1}$ because it is an element of Γ_1 and hence acts trivially.

To show that $f|[\Gamma_1\alpha\Gamma_2]_k$ is a modular form of weight k for Γ_2 we need to show that it transforms correctly under the action of Γ_2. This follows because multiplication by an element $g_2 \in \Gamma_2$ on the right permutes the orbit space $\Gamma_1\backslash\Gamma_1\alpha\Gamma_2$ by right multiplication. So if $\{\beta_j\}$ is a set of coset representatives, then $\{\beta_j g_2\}$ is also a set of orbit representatives. This means that

$$(f|[\Gamma_1\alpha\Gamma_2]_k)|[g_2]_k = \sum_j f|[\beta_j g_2]_k = f|[\Gamma_1\alpha\Gamma_2]_k.$$

This shows that $f|[\Gamma_1\alpha\Gamma_2]_k$ is *weakly modular*; we now need to show that it is holomorphic at the cusps; this follows because we end up considering a sum of functions, each of which we can show to be holomorphic at the cusps. This sum of functions is itself holomorphic at the cusps.

Finally, if f is a cusp form then it can be shown that the sum of functions in the previous paragraph is a sum of holomorphic functions, each of which vanishes at all of the cusps, so therefore $f|[\Gamma_1\alpha\Gamma_2]_k$ is a cusp form if f is a cusp form.

Although we have outlined this in a great deal of generality, in practice we are very interested in only a few of these operators. In [Diamond and Shurman (2005)], it is shown that every double coset operator is given by a combination of the three special cases below:

(1) $\Gamma_1 \subset \Gamma_2$. This gives a trace map which projects $M_k(\Gamma_1)$ onto $M_k(\Gamma_2)$.
(2) $\Gamma_1 = \alpha^{-1} \cdot \Gamma_2 \cdot \alpha$. Here, the double coset operator map gives an isomorphism between $M_k(\Gamma_1)$ and $M_k(\Gamma_2)$.
(3) $\Gamma_1 \supset \Gamma_2$. A modular form for Γ_1 will also be a modular form for Γ_2, because it has to satisfy fewer conditions; the double coset operator becomes the inclusion map from $M_k(\Gamma_1)$ into $M_k(\Gamma_2)$.

There are two types of "standard" Hecke operators acting on modular forms of level $\Gamma_1(N)$ that one encounters in nature. These are the Hecke operators T_n, which generalize the operators T_p introduced in the previous section, and the *diamond operators*, which we define here.

Definition 4.4. Let N be a positive integer and let $f \in M_k(\Gamma_1(N))$. Let $\alpha = \begin{pmatrix} a & b \\ c & d \end{pmatrix} \in \Gamma_0(N)$. We define the *diamond operator* $\langle d \rangle$ to be the double coset operator $f|[\Gamma_1(N)\alpha\Gamma_1(N)]_k$.

Because $\Gamma_1(N)$ is a normal subgroup of $\Gamma_0(N)$, the diamond operator is number (2) in the list above. The action of α is completely determined by d modulo N because the subgroup $\Gamma_1(N)$ acts trivially on f.

If $f \in M_k(\Gamma_0(N), \chi)$, then it can be shown that the diamond operator acts on f as multiplication by $\chi(d)$.

We will now define the Hecke operators T_p for modular forms of level $\Gamma_1(N)$ in terms of the double coset operators. It can also be shown that the diamond operators commute with the T_p.

Theorem 4.5. *Let N be a positive integer and let p be a prime number. The action of the double coset operator $[\Gamma_1(N)\left(\begin{smallmatrix}1&0\\0&p\end{smallmatrix}\right)\Gamma_1(N)]_k$ on $M_k(\Gamma_1(N))$ is given by*

$$T_p(f) = \begin{cases} \sum_{j=0}^{p-1} f|[\begin{pmatrix}1&j\\0&p\end{pmatrix}]_k, & \text{if } p|N, \\ \sum_{j=0}^{p-1} f|[\begin{pmatrix}1&j\\0&p\end{pmatrix}]_k + f|[\left(\begin{smallmatrix}m&n\\N&p\end{smallmatrix}\right)\cdot\left(\begin{smallmatrix}p&0\\0&1\end{smallmatrix}\right)]_k, & \text{if } p\nmid N, \end{cases} \tag{4.6}$$

where m, n are integers with $mp - Nn = 1$.

Proof. We refer to Section 5.2 of [Diamond and Shurman (2005)] for the details of this proof, which relies on a careful computation of the double coset, and then an analysis which gives the explicit representation of the action of the double coset. □

Remark 4.6. If we change $\Gamma_1(N)$ to $\Gamma_0(N)$ in Theorem 4.5, and keep α the same, then we can replace the final matrix in (4.6) with $\left(\begin{smallmatrix}p&0\\0&1\end{smallmatrix}\right)$, because the matrix $\gamma := \left(\begin{smallmatrix}m&n\\N&p\end{smallmatrix}\right)$ is an element of $\Gamma_0(N)$, and therefore $f|[\gamma]_k = f$.

Having found out how these Hecke operators act on spaces of modular forms, we now consider what their effect is on Fourier expansions.

Theorem 4.7. *Let f in $M_k(\Gamma_0(N), \chi)$ be a modular form with Fourier expansion $f(q) = \sum_{n=1}^{\infty} a_n q^n$ and let m be a positive integer. The action of the Hecke operator T_m on the Fourier expansion of f is given by*

$$(T_m f)(q) = \sum_{n=0}^{\infty} b_n q^n, \tag{4.7}$$

where

$$b_n = \sum_{1 \leq d | \gcd(m,n)} \chi(d) d^{k-1} a_{mn/d^2}.$$

Remark 4.8. We note that this proof generalizes that of Theorem 4.2. A useful point to note is that it is much easier to compute the Fourier expansion of $f\left((az+b)/(cz+d)\right)$ if $c = 0$.

Proof. We will first consider $n = p$ where p is prime. We look at exactly how the matrices given in the definition of the Hecke operator act on Fourier expansions. Let $0 \le j < p$; we see that

$$f| \left[\begin{pmatrix} 1 & j \\ 0 & p \end{pmatrix} \right]_k (z) = p^{k-1}(0z+p)^{-k} \cdot f\left(\frac{z+j}{p}\right) = \frac{1}{p} \cdot \sum_{n=0}^{\infty} a_n(f) e^{2\pi i n(z+j)/p}$$

$$= \frac{1}{p} \sum_{n=0}^{\infty} a_n(f) q_p^n \cdot \zeta_p^{nj},$$

where $q_p = e^{2\pi i z/p}$ and $\zeta_p = e^{2\pi i/p}$ is a pth root of unity. It is a well-known fact from number theory that the sum $\sum_{j=0}^{p-1} \zeta_p^{nj}$ is p if $p|N$ and 0 otherwise, so we find that, if we sum over j then

$$\sum_{j=0}^{p-1} f| \left[\begin{pmatrix} 1 & j \\ 0 & p \end{pmatrix} \right]_k (z) = \sum_{\substack{n=0 \\ n \equiv 0 \mod p}}^{\infty} a_n(f) q_p^n = \sum_{n=0}^{\infty} a_{np}(f) q^n.$$

If $p \nmid N$, then we have one extra term, which gives the additional summand

$$f|[(\begin{smallmatrix} m & n \\ N & p \end{smallmatrix}) \cdot (\begin{smallmatrix} p & 0 \\ 0 & 1 \end{smallmatrix})]_k(z) = (\langle p \rangle f)[(\begin{smallmatrix} p & 0 \\ 0 & 1 \end{smallmatrix})]_k(z)$$

$$= p^{k-1}(0z+1)^{-k}(\langle p \rangle f)(pz) = p^{k-1} \sum_{n=0}^{\infty} a_n(\langle p \rangle f) q^{np};$$

(the action of the matrix $(\begin{smallmatrix} m & n \\ N & p \end{smallmatrix})$ is trivial because it is an element of $\Gamma_0(N)$), and the action of $\langle p \rangle$, as we have seen above, is to multiply by the character evaluated at p. This gives the Fourier expansion for $n = p$.

We now let n be an arbitrary positive composite integer. We define the Hecke operators for prime powers in a similar way to the level 1 case, by

$$T_{p^{r+1}} = T_p \cdot T_{p^r} - p^{k-1} \langle p \rangle \cdot T_{p^{r-1}}, \text{ if } r \ge 1, \tag{4.8}$$

and we define $T_{mn} := T_m \cdot T_n$ if $(m, n) = 1$. It can be shown that the formula in the theorem holds for p^r and then for n by induction on r. □

We note that if $n = p$ is prime then (4.7) simplifies to

$$(T_p f)(q) = \sum_{n=1}^{\infty} \left(a_{pn} + \chi(p) p^{k-1} a_{n/p} \right) q^n,$$

where we have the convention that if n/p is not an integer then $a_{n/p}$ is zero, and if $p|N$ then $\chi(p) = 0$. Also, if we take $\chi = 1$ and let $N = 1$, we see that we recover the formula given earlier for level 1; this shows that the different definitions of Hecke operators are in fact compatible and give the same answers.

We have that, if $f \in M_k(\Gamma_0(N), \chi)$, then equation (4.8) becomes

$$T_{p^{r+1}} = T_{p^r} \cdot T_p - \chi(p) \cdot p^{k-1} \cdot T_{p^{r-1}}, \tag{4.9}$$

as the diamond operator $\langle p \rangle$ acts as multiplication by $\chi(p)$ here.

The algebra generated by these Hecke operators over a given ring is of independent arithmetic interest: we define it here.

Definition 4.9. Let Γ be a congruence subgroup and let R be a ring. We define $\mathbf{T}_R(S_k(\Gamma))$ to be the $\mathbf{Z}$-algebra generated by the Hecke operators T_n acting on $S_k(\Gamma)$, tensored over $\mathbf{Z}$ with R. This is known as the *Hecke algebra associated to* $S_k(\Gamma)$.

Some authors consider the Hecke algebra generated by only those T_n which have $(n, N) = 1$, where N is the level of Γ. This is sometimes called the *anaemic Hecke algebra.*

We will see later that this is in fact a finitely generated algebra over R (standard choices for R are $\mathbf{Z}$ or a finite field $\mathbf{F}_q$), and in Exercise 6 the reader can compute a Hecke algebra with small dimension for themselves. In Chapter 7 we will see how to compute Hecke algebras with computer algebra packages.

In Section 3.3, we saw that there is an explicit bound, called the Sturm bound, on how many coefficients of a modular form we need to compute to uniquely identify it. The paper [Kilford and Wiese (2008)] considers the question of the Sturm bound's sharpness for computing Hecke *algebras* (especially over a field), and gives algorithms which can terminate before reaching the Sturm bound if the entire algebra has been computed.

We can use the Sturm bound to compute the number of Hecke operators needed to generate Hecke algebras; there is an interesting remark (Remark 9.24) in [Stein (2007)] which says that it does not suffice in general to use only the operators T_p where p is *prime* and less than or equal to the Sturm bound to compute the Hecke algebra as a ring over $\mathbf{Z}$. For instance, let us consider the one-dimensional space $S_4(\Gamma_0(5))$. We can check that the Sturm bound here is 2, and the eigenvalue of T_2 acting on this space is -4, so the algebra $\mathbf{Z}[T_2]$ is $4\mathbf{Z}$. However, $\mathbf{T}_\mathbf{Z}(S_4(\Gamma_0(5))) = \mathbf{Z}$; we need to also take into account the Hecke operator T_1.

There are also examples of this phenomenon where the dimension of the space of modular forms is greater than 1; in Exercise 5, we will see how to find such spaces computationally.

We will now say a few words about the arithmetic interest of Hecke algebras. The most famous application of these is undoubtedly in the proof by Taylor and Wiles of Fermat's Last Theorem; we note that [Taylor and Wiles (1995)] is about ring-theoretic properties of Hecke algebras.

However, they do have other applications. In a slightly broader context, they have been used by [Lansky and Pollack (2002)], to prove that, if a certain conjecture of Gross holds, then there is a Galois extension of $\mathbf{Q}$ with Galois group $\mathrm{GL}_2(\mathbf{F}_5)$ which is ramified only at 5.

The paper [Calegari and Stein (2004)] considers them for their arithmetic interest; it computes many discriminants of Hecke algebras, and gives conjectures which relate the p-divisibility of these discriminants to the existence or otherwise of mod p congruences between modular forms in $S_k(\Gamma_0(p))$. One of the conjectures of this paper has been proved recently in [Ahlgren and Barcau (2007)].

The book [Krieg (1990)] gives an overview of Hecke algebras in a wider context.

Remark 4.10. It is very important to note that the action of the Hecke operator T_n on a modular form f of level N which is prime to n is different to the action of the Hecke operator T_n on the modular form f considered as a modular form of level Nn. This illustrates that we have to remember what the level of a form is; we *cannot* simply look at the Fourier expansion at ∞. It is important to remember that the modular form is not just its Fourier expansion.

As an example of this, let us consider the modular form $f := \Delta \cdot E_4 \in S_{16}(\mathrm{SL}_2(\mathbf{Z}))$. We will see later that this is a simultaneous eigenform for all of the Hecke operators T_n as it is the unique normalized cuspform of weight 16 for $\mathrm{SL}_2(\mathbf{Z})$.

However, we can also view it as a modular form of weight 16 and level $\Gamma_0(2)$ via the inclusion map $\imath : S_k(\mathrm{SL}_2(\mathbf{Z})) \hookrightarrow S_k(\Gamma_0(2))$. The form $\imath(f)$ now has the trivial character modulo 2, so we see that the action of T_2 is now given by

$$(T_2\imath(f))(q) = \sum_{n=1}^{\infty} a_{2n} q^n.$$

Now f is *not* an eigenform for this new T_2 (we can check this explicitly, or note that it is an eigenform for the normal T_2 and that the extra part of the normal T_2 is nonzero). We can find eigenvectors for this new T_2 by considering its action on the space consisting of $\imath(f)(q)$ and $f(q^2)$ (see Exercise 22).

For this reason, many authors use the following notation for what we have been calling the "new T_2":

Definition 4.11. Let $M := M_k(\Gamma_0(N), \chi)$ be a space of modular forms. If $p|N$, then we call the Hecke operator T_p acting on M U_p; its effect on

Fourier coefficients is

$$(U_p f)(q) = \sum_{n=1}^{\infty} a_{pn} q^n.$$

Another standard notation is to call the operator mapping $M_k(\Gamma_1(N))$ into $M_k(\Gamma_1(Np))$ whose action on Fourier expansions is to send q to q^p either V or V_p. We see that the only forms which are eigenvalues for V_p are the constants (because the rings of power series we are dealing with are not infinitely divisible).

4.2 Bases of eigenforms

In Chapter 3, we found a basis for the space of modular forms of level 1 in terms of the Eisenstein series E_4 and E_6. However, this basis is not the most useful choice; for instance, it is not composed of eigenforms for the Hecke operators T_p in general (see [Emmons (2005)] for a proof of this). We have already seen bases of eigenforms for some small-dimensional spaces of modular forms; for instance, $M_{12}(\mathrm{SL}_2(\mathbf{Z}))$ has the basis $\{E_{12}, \Delta\}$, and both of these forms are eigenforms for the Hecke operators (as can be seen from dimension considerations).

In this section, we will prove that bases of eigenforms for all the T_p exist, under certain conditions, and we will investigate those cases where we cannot find bases of eigenforms for all the T_p.

4.2.1 *The Petersson scalar product*

We define the *Petersson scalar product* on modular forms.

Definition 4.12 (Petersson scalar product). *Let Γ be a congruence subgroup, let k be an integer and let F be a fundamental domain for Γ. Let $f, g \in M_k(\Gamma)$, with at least one of f and g being a cuspidal modular form, and let $\overline{g}$ be the complex conjugate of g. We define the* Petersson scalar product *or* Petersson inner product *to be*

$$\langle f, g \rangle := \frac{1}{[\mathrm{PSL}_2(\mathbf{Z}) : \mathrm{PSL}_2(\mathbf{Z}) \cap \Gamma]} \int_F f(z)\overline{g(z)} y^k \frac{dxdy}{y^2}.$$

We note some properties of this inner product which will be useful. It is clear from the definition that it is linear in the first factor f and antilinear in the second factor g; by this, we mean that $\langle f, c \cdot g \rangle = \overline{c} \cdot \langle f, g \rangle$, for $c \in \mathbf{C}$

(where again $\overline{x}$ means the complex conjugate of x). It is also antisymmetric; we see that $\langle f, g\rangle = \overline{\langle g, f\rangle}$ straight from the definition, and finally we see that it is also positive definite; we have $\langle f, f\rangle > 0$, from a consideration of the product inside the integral, and the knowledge that y is positive in the fundamental domain F (because $F \subset \mathcal{H}$). This means that the Petersson scalar product is *Hermitian.*

We require at least one of f and g to be cuspidal to ensure that the integral converges; this ensures that the integral vanishes at all of the cusps.

Remark 4.13. The Eisenstein series E_k are orthogonal to the cusp forms $S_k(\Gamma)$; that is, $\langle E_k, f\rangle = \langle f, E_k\rangle = 0$ for every $f \in S_k(\Gamma)$. In fact, one way to *define* Eisenstein series in general is to say that they are the space of forms which are orthogonal to every cusp form with respect to the Petersson scalar product. Following the discussion at the end of Section III.5 of [Koblitz (1993)], this is an intrinsic way to define the Eisenstein series, without reference to a specific congruence subgroup.

The multiplier before the integral in the definition of the inner product is there to ensure that the scalar product is independent of the choice of Γ; it is for this reason that we don't need to carry around a subscript on the scalar product to keep track of the group that f and g are defined over. This is in contrast to the situation with Hecke operators, where they do depend on the level of the modular forms they operate on. We note that some authors do define the inner product without this normalization.

Now we will show that $\langle , \rangle$ is well-defined; in particular, we need to prove our claim that it is independent of the choice of fundamental domain F for Γ.

Lemma 4.14. *The Petersson scalar product $\langle , \rangle$ is well-defined, and independent of the choice of fundamental domain F. Also, if $f, g \in M_k(\Gamma_0(N), \chi)$ (with at least one of f and g being a cusp form) and $\alpha \in \mathrm{GL}_2^+(\mathbf{Q})$ then*

$$\langle f, g\rangle = \langle f|[\alpha]_k, g|[\alpha]_k\rangle \text{ and } \langle f|[\alpha]_k, g\rangle = \langle f, g|[\det(\alpha)\cdot\alpha^{-1}]_k\rangle. \quad (4.10)$$

Also, $\langle f|[\alpha]_k, g\rangle$ is only dependent on the double coset $\Gamma\alpha\Gamma$.

Proof. We will define the *hyperbolic measure* on the Poincaré upper half plane $\mathcal{H}$ by

$$d\mu(z) = \frac{dxdy}{y^2}, \text{ where } z = x + iy \in \mathcal{H}.$$

We follow Koblitz's exposition in Section III.5 of [Koblitz (1993)] in the proof of this lemma. First, we will show that $\int_F d\mu(z)$ converges; firstly, we see that if $\Gamma = \mathrm{SL}_2(\mathbf{Z})$ then

$$\int_F d\mu(z) = \int_F \frac{dxdy}{y^2} < \int_{-1/2}^{1/2} \int_{\sqrt{3}/2}^{\infty} y^{-2} dy dx = \frac{2}{\sqrt{3}}. \tag{4.11}$$

In the second integral we are using the fact that the standard fundamental domain for $\mathrm{SL}_2(\mathbf{Z})$ is the one given in Figure 2.1. Checking the inequalities here and computing the integral is left as Exercise 8.

By Exercise 17, we see that this hyperbolic measure is invariant under the action of $\mathrm{SL}_2(\mathbf{Z})$, and in fact under the action of $\mathrm{GL}_2^+(\mathbf{Q})$. If we were to choose a new fundamental domain G for Γ, then we could partition G into regions R such that there exists $\alpha \in \mathbf{F}$ with $\alpha R \subset F$, and then we use the result that $d\mu(z)$ is invariant under the Möbius transformation $z \mapsto \alpha z$ to show that the integrals over R as a region of G and over αR as a region of F are equal. This shows that $\langle , \rangle$ is independent of the choice of fundamental domain, and also that $d\mu(z)$ is finite for any choice of Γ.

We will now prove the identities given in (4.10) by explicitly considering the definition of the scalar product. Let $\Gamma_2 := \Gamma \cap \alpha^{-1}\Gamma\alpha$; this is a congruence subgroup, and we know that $f|[\alpha]_k$ and $g|[\alpha]_k$ are modular forms in $M_k(\Gamma_2)$, by the classification of Hecke operators given earlier. We have that

$$\begin{aligned}
&\langle f|[\alpha]_k, g|[\alpha]_k \rangle && (4.12)\\
&= \frac{1}{[\mathrm{PSL}_2(\mathbf{Z}) : \mathrm{PSL}_2(\mathbf{Z}) \cap \Gamma]} \int_{\alpha^{-1}F} f|[\alpha]_k(z)\overline{g(z)|[\alpha]_k}\frac{dxdy}{y^2} && (4.13)\\
&= \frac{1}{[\mathrm{PSL}_2(\mathbf{Z}) : \mathrm{PSL}_2(\mathbf{Z}) \cap (\alpha^{-1}\Gamma_2\alpha)]} \int_{F_2} f(z)\overline{g(z)}y^k\frac{dxdy}{y^2} && (4.14)\\
&= \frac{1}{[\mathrm{PSL}_2(\mathbf{Z}) : \mathrm{PSL}_2(\mathbf{Z}) \cap \Gamma_2]} \int_{F_2} f(z)\overline{g(z)}y^k\frac{dxdy}{y^2} && (4.15)\\
&= \langle f, g \rangle. && (4.16)
\end{aligned}$$

In equation (4.12), we are using the fact that the hyperbolic measure is invariant under the Möbius transformation $z \mapsto \alpha z$), and in equation (4.15), we are using the fact that the congruence subgroups are isomorphic (under the action of conjugation), so they have the same index inside $\mathrm{PSL}_2(\mathbf{Z})$.

Now we will show that $\langle f|[\alpha]_k, g\rangle = \langle f, g|[\alpha^{-1}]_k\rangle$. We define $\alpha_2 := \det(\alpha)\cdot\alpha^{-1}$; we have that the weight-k action of α_2 and of α^{-1} is the same, as multiplication by scalars is a trivial action there. We now replace g

by $g|[\alpha^{-1}]_k$, and change our congruence subgroup to be $\alpha\Gamma\alpha^{-1}$, in the first identity in (4.10), which we have just proved, to obtain

$$\langle f|[\alpha]_k, g\rangle = \langle f, g|[\alpha^{-1}]_k\rangle,$$

as required.

Finally, we now show that $\langle f|[\alpha], g\rangle$ depends only on the choice of an element in the double coset $\Gamma\alpha\Gamma$. Let $\gamma_1, \gamma_2 \in \Gamma$. We now perform the following computations:

$$\langle f|[\gamma_1\alpha\gamma_2]_k, g\rangle = \langle f|[\gamma_1\alpha]_k, g|[\gamma_2^{-1}]_k\rangle \tag{4.17}$$

$$= \langle f|[\alpha]_k, g\rangle; \tag{4.18}$$

we apply the results we have just proved, to move the $[\gamma_2]_k$ from one form to the other, and also use the fact that f and g are modular forms and are therefore weight-k invariant under the action of elements of Γ. □

We have now shown that the complex vector space $S_k(\Gamma_0(N), \chi)$, equipped with the Petersson scalar product, is an inner product space; it is in fact a Hilbert space[1].

Now we will prove a lemma which will give us a very useful set of coset representatives for the double coset $\Gamma\alpha\Gamma$. We follow the account given in [Diamond and Shurman (2005)], Section 5.5.

Lemma 4.15. *Let $\alpha \in \mathrm{GL}_2^+(\mathbf{Q})$ and let Γ be a congruence subgroup. There exists a set $\{\beta_j\}$ of elements of $\mathrm{GL}_2^+(\mathbf{Q})$ such that*

$$\Gamma\alpha\Gamma = \bigcup \Gamma\beta_j = \bigcup \beta_j\Gamma.$$

Proof. There exist $\{\gamma_i\}$ and $\{\delta_j\}$ in Γ such that

$$\Gamma = \bigcup_j (\alpha^{-1}\Gamma\alpha \cap \Gamma)\gamma_j = \bigcup_j (\alpha^{-1}\Gamma\alpha \cap \Gamma)\delta_j^{-1},$$

with both of these unions being disjoint. We have the following representatives in the double cosets:

$$\Gamma\alpha\Gamma = \bigcup_j \Gamma\alpha\gamma_j \text{ and } \Gamma\alpha^{-1}\Gamma = \bigcup_j \Gamma\alpha^{-1}\delta_j^{-1};$$

these are disjoint unions. We can rewrite the second of these as

$$\Gamma\alpha\Gamma = \bigcup_j \delta_j\alpha\Gamma.$$

[1] It is said that David Hilbert and Richard Courant once attended a lecture which discussed Hilbert spaces. After this lecture, Hilbert supposedly turned to Courant and asked "Richard, exactly what is a Hilbert space?"

We can further show that the cosets $\Gamma\alpha\gamma_i$ and $\delta_j\alpha\Gamma$ have nontrivial intersection, as otherwise we have $\Gamma\alpha\gamma_j \subset \bigcup_{i\neq j}\delta_i\alpha\Gamma$, and if we multiply on the right by Γ then we can see that the double coset $\Gamma\alpha\Gamma$ is a subset of $\bigcup_{i\neq j}\delta_i\alpha\Gamma$, which is false. Therefore, for every j we can choose an element of the nonempty set $\Gamma\alpha\gamma_j \cap \delta_j\alpha\Gamma$ and this will be β_j, with the properties we require of it.

We also note for reference in the next theorem that if we have a coset decomposition of $\Gamma\alpha\Gamma = \bigcup_j \Gamma\beta_j$, then we also have the decomposition

$$\Gamma\left(\det(\alpha)\alpha^{-1}\right)\Gamma = \bigcup_j \Gamma\left(\det(\beta_j)\beta_j^{-1}\right). \tag{4.19}$$

This follows by manipulation of the cosets. □

We will now use the results we have just derived to find out exactly how the Petersson scalar product and the Hecke operators interact.

Theorem 4.16. *Let N be a positive integer, and let $\chi : (\mathbf{Z}/N\mathbf{Z})^\times \to \mathbf{C}^\times$ be a Dirichlet character. Let n be a positive integer with $(n, N) = 1$.*

If $f, g \in M_k(\Gamma_0(N), \chi)$ (with at least one of f and g being a cusp form) then we have that

$$\langle T_n f, g\rangle = \chi(n)\langle f, T_n g\rangle. \tag{4.20}$$

Proof. If we can show that this holds for all T_p where p is prime (and $p \nmid N$), then we can use the formulae given for the Hecke operators T_n to prove the theorem for all of the T_n. We recall from Theorem 4.5 that we have the following identity for $T_p f$:

$$\begin{aligned} T_p f &= f|[\Gamma_1(N)\left(\begin{smallmatrix}1&0\\0&p\end{smallmatrix}\right)\Gamma_1(N)]_k \\ &= \sum_{j=0}^{p} f|[\beta_j]_k \end{aligned}$$

where the β_j here are obtained using Lemma 4.15, so they commute with $\Gamma_1(N)$. We note that all of these β_j will have determinant p.

We now use this for the final calculation:

$$\begin{aligned}
\langle T_p f, g\rangle &= \langle f|[\Gamma_1(N)\left(\begin{smallmatrix}1&0\\0&p\end{smallmatrix}\right)\Gamma_1(N)]_k, g\rangle && (4.21)\\
&= \left\langle \sum_{i=0}^{p} f|[\beta_i]_k, g\right\rangle \text{ (by the definition of } T_p) && (4.22)\\
&= \sum_{i=0}^{p} \langle f|[\beta_i]_k, g\rangle \text{ (by linearity)} && (4.23)\\
&= \sum_{i=0}^{p} \langle f, g|[\det(\beta)\beta_i^{-1}]_k\rangle \text{ (from (4.12))} && (4.24)\\
&= \left\langle f, \sum_{i=0}^{p} g|[\det(\beta)\beta_i^{-1}]_k\right\rangle \text{ (from (4.19))} && (4.25)\\
&= \langle f, g|[\Gamma_1(N)\left(\begin{smallmatrix}p&0\\0&1\end{smallmatrix}\right)\Gamma_1(N)]_k\rangle, \text{ (by linearity).} && (4.26)
\end{aligned}$$

We now have to deal with this new Hecke operator. We note that the β_j have been chosen using Lemma 4.15.

We can choose integers m and n such that $mp - nN = 1$, and by direct computation, we see that

$$\left(\begin{smallmatrix}p&0\\0&1\end{smallmatrix}\right) = \left(\begin{smallmatrix}1&n\\N&mp\end{smallmatrix}\right)^{-1} \cdot \left(\begin{smallmatrix}1&0\\0&p\end{smallmatrix}\right) \cdot \left(\begin{smallmatrix}p&n\\N&m\end{smallmatrix}\right).$$

We see that the first matrix in the product is in $\Gamma_1(N)$ and the third is an element of $\Gamma_0(N)$. This means that we have the following decomposition of $\Gamma_1(N)\left(\begin{smallmatrix}p&0\\0&1\end{smallmatrix}\right)\Gamma_1(N)$:

$$\Gamma_1(N)\left(\begin{smallmatrix}p&0\\0&1\end{smallmatrix}\right)\Gamma_1(N) = \Gamma_1(N)\left(\begin{smallmatrix}1&0\\0&p\end{smallmatrix}\right)\Gamma_1(N)\left(\begin{smallmatrix}p&n\\N&m\end{smallmatrix}\right); \quad (4.27)$$

we obtain this by substituting in for $\left(\begin{smallmatrix}p&0\\0&1\end{smallmatrix}\right)$ and then using the fact that $\Gamma_1(N)$ is a *normal* subgroup of $\Gamma_0(N)$.

This relates the coset representatives for T_p and for the new Hecke operator that we obtained in (4.26); we find that if we have the decomposition

$$T_p = [\Gamma_1(N)\left(\begin{smallmatrix}1&0\\0&p\end{smallmatrix}\right)\Gamma_1(N)]_k = \bigcup_j \Gamma_1(N)\beta_j$$

then the corresponding decomposition for $[\Gamma_1(N)\left(\begin{smallmatrix}p&0\\0&1\end{smallmatrix}\right)\Gamma_1(N)]_k$ is

$$[\Gamma_1(N)\left(\begin{smallmatrix}p&0\\0&1\end{smallmatrix}\right)\Gamma_1(N)]_k = \bigcup_j \Gamma_1(N)\beta_j\left(\begin{smallmatrix}p&n\\N&m\end{smallmatrix}\right).$$

We now use the fact that $m \equiv p^{-1} \mod N$ to see that the matrix $R := \left(\begin{smallmatrix}p&n\\N&m\end{smallmatrix}\right)$ acts as the diamond operator $\langle p^{-1}\rangle$, which acts by multiplication by $\chi(p^{-1})$, so we have

$$\begin{aligned}\langle f, g|[\Gamma_1(N)\left(\begin{smallmatrix} p & 0 \\ 0 & 1\end{smallmatrix}\right)\Gamma_1(N)]_k\rangle &= \left\langle f, \sum_{j=0}^{p} g|[\beta_j \cdot R]_k \right\rangle \\ &= \left\langle f|[R^{-1}]_k, \sum_{j=0}^{p} g|[\beta_j]_k \right\rangle \text{ (by (4.10))} \\ &= \chi(p) \cdot \langle f, T_p g\rangle ,\end{aligned}$$

as required. (It is easy to show that the bottom right coefficient of R^{-1} is congruent to p modulo N). □

4.2.2 *The Hecke operators are Hermitian*

We now use this result to show that the Hecke operators are Hermitian with respect to the Petersson scalar product.

Theorem 4.17. *Let $f, g \in M_k(\Gamma_0(N), \chi)$ be modular forms, at least one of which is a cusp form. The Petersson scalar product $\langle f, g\rangle$ is Hermitian and non-degenerate.*

Let n be a positive integer which is prime to N, and let c_n be a square root of $\chi(n)$. Then we have

$$\langle c_n T_n f, g\rangle = \langle f, c_n T_n g\rangle;$$

in other words, if $(n, N) = 1$ then the Hecke operators T_n are Hermitian operators with respect to the Petersson scalar product.

Proof. We have shown that the scalar product is well-defined and non-degenerate. It now remains to check that that it satisfies the equation given above, which we do directly. (Here, $\overline{\alpha}$ is the complex conjugate of α).

$$\begin{aligned}\langle c_n T_n f, g\rangle &= c_n \langle T_n f, g\rangle \text{ (linear in } f) \\ &= c_n \cdot \chi(n)\langle f, T_n g\rangle \text{ (from Theorem 4.16)} \\ &= c_n \overline{c}_n^2 \langle f, T_n g\rangle \text{ (by the definition of } \chi) \\ &= \overline{c}_n \langle f, T_n g\rangle \\ &= \langle f, c_n T_n g\rangle\end{aligned}$$

as required. □

We will show that there are specific spaces of modular forms called *new-forms* $M_k^{\text{new}}(\Gamma)$ and $S_k^{\text{new}}(\Gamma)$; these have bases consisting of eigenforms for all of the Hecke operators. In contrast, if there is a sufficiently large power of a prime dividing the level, it is possible to show that there is no basis of forms which are eigenvalues for all of the Hecke operators.

The following theorem on the orthogonality of cusp forms with different Dirichlet characters can be found in [Lang (1976)] as Theorem VII.5.2; we leave the proof to the reader as Exercise 7:

Theorem 4.18. *Let N be a positive integer and let $\chi, \mu : (\mathbf{Z}/N\mathbf{Z})^\times \to \mathbf{C}^\times$ be distinct Dirichlet characters. If $f \in S_k(\Gamma_0(N), \chi)$ and $g \in S_k(\Gamma_0(N), \mu)$, then we have*

$$\langle f, g \rangle = 0.$$

Proof. See Exercise 7. □

The following theorem is called the *Spectral Theorem of linear algebra*; if we have a finite-dimensional vector space with a set of operators acting on it which commute with each other, then we have an orthogonal basis for the space.

Theorem 4.19. *Let V be a finite-dimensional complex vector space equipped with a positive definite Hermitian form $\langle , \rangle$.*

(1) Let $t : V \to V$ be a linear map which is Hermitian; in other words, if $v, w \in V$ then $\langle t(v), w \rangle = \langle v, t(w) \rangle$. Then V has a basis consisting of eigenvectors for t (so t is diagonalizable).

(2) Let $t_1, t_2, \ldots$ be a sequence of Hermitian operators sending V to V which commute with each other. Then V has a basis consisting of vectors which are eigenvectors for all *of the t_i (so the t_i are* simultaneously *diagonalizable).*

Proof.

(1) Because $\mathbf{C}$ is algebraically closed, we see that t has an eigenvector v_1 by considering the characteristic polynomial of t acting on V and noting that it must have at least one root. We now define $V_2 := (\mathbf{C}v_1)^\perp$. As t is a Hermitian map, V_2 is stable under the action of t, so by the same argument t has an eigenvector v_2 in V_2. We define $V_3 := (\mathbf{C}v_1 + \mathbf{C}v_2)^\perp$, and continue in the same way. Because V is finite-dimensional, and we are dealing with strictly smaller vector spaces at each step, this process must terminate, and so we are done.

(2) We know from (1) that we can write $V = \oplus_i V_i$, where each V_i is the subspace of V where t_1 acts as multiplication by λ_i, with each λ_i distinct. We now use the fact that t_2 commutes with t_1; this means that t_2

preserves each of the spaces V_i, so we can use the arguments from the proof of (1) to decompose V_i into a sum of eigenspaces for t_2. We repeat this process for each successive t_j until we obtain a decomposition $V = \oplus V_k$ where all of the t_j act as scalars on all of the V_k. We now choose bases for the V_k and take the union of these bases; this is a basis of simultaneous eigenvectors for all the t_j. □

Remark 4.20. We see that in this proof we have used that $\mathbf{C}$ is algebraically closed, to show that a polynomial over $\mathbf{C}$ has to have a root, and the fact that V is finite-dimensional, to ensure that this process will end. For the computational issues we are interested in, we note that the field $\overline{\mathbf{Q}}$ would have sufficed, and in fact after we have performed our calculations we can choose a finite extension K of $\mathbf{Q}$ such that our operators t have a root in K.

We can now use the results of this section to prove the following very useful theorem on bases of eigenforms for almost all the Hecke operators T_p.

Theorem 4.21. *Let N be a positive integer and let $\chi : (\mathbf{Z}/N\mathbf{Z})^\times \to \mathbf{C}^\times$ be a Dirichlet character. Then the space $S_k(\Gamma_0(N), \chi)$, viewed as a complex vector space, has a basis whose elements are eigenforms for all of the Hecke operators T_n such that $(n, N) = 1$.*

Proof. From Theorem 4.17, we see that if $(n, N) = 1$ then T_n is a Hermitian operator with respect to the Petersson scalar product, and therefore we can apply Theorem 4.19 to the set of T_n which have $(n, N) = 1$. □

Remark 4.22. It can be seen that this is best possible in the following sense; there exist spaces of modular forms where one does *not* have a basis of simultaneous eigenforms for all the T_n (there exist integers n dividing N such that there is no basis of eigenforms for T_n); this is explored in Exercise 3. Briefly, the problem that this exercise illustrates is that if there are "sufficiently many" oldforms present then some of the T_n will fail to be Hermitian.

We can also show that the coefficients of a normalized eigenform for $\mathrm{SL}_2(\mathbf{Z})$ must be real numbers (of course, if we allow the eigenform to be non-normalized, we cannot expect the coefficients to be real).

Corollary 4.23. *Let f be a normalized cuspidal eigenform for $\mathrm{SL}_2(\mathbf{Z})$ of weight k, with Fourier expansion $\sum_{n=0}^{\infty} a_n q^n$. Then the a_n are real.*

Proof. If f is *not* cuspidal and normalized, then it must be the Eisenstein series E_k, which is an eigenform with rational coefficients. Therefore, without loss of generality we may assume that f is a nonzero cusp form.

Let T_n be a Hecke operator with $T(f) = af$. Then we have

$$a \cdot \langle f, f\rangle = \langle af, f\rangle = \langle T(f), f\rangle \tag{4.28}$$
$$= \langle f, T_n(f)\rangle = \langle f, af\rangle = \overline{a} \cdot \langle f, f\rangle \tag{4.29}$$

Because the Petersson scalar product is positive definite and $f \neq 0$, $\langle f, f\rangle \neq 0$, so a must be its own complex conjugate, so it must be real. This holds for all of the Hecke operators T_n, so all of the a_n are real numbers, as required. □

Let us consider an example where we can find a basis of eigenforms. We can check that the space of modular forms $S_{28}(\mathrm{SL}_2(\mathbf{Z}))$ is 2-dimensional; we can generate it by $f_1 := \Delta \cdot E_4^4$ and $f_2 := \Delta^2 \cdot E_4$; these are both cusp forms, as they vanish at ∞, and they have weight 28, because they are the product of modular forms whose weights sum to 28. They are linearly independent, because $\nu_\infty(f_1) = 1$ and $\nu_\infty(f_2) = 2$. They have Fourier expansions

$$f_1(q) = q + 936q^2 + 331452q^3 + 53282368q^4 + O(q^5) \tag{4.30}$$
$$f_2(q) = q^2 + 192q^3 - 8280q^4 + O(q^5). \tag{4.31}$$

We will consider the action of the Hecke operator T_2 on these forms; we find that

$$T_2(f_1) = 936q + 187500096q^2 + O(q^3) = 936f_1 + 18662400f_2 \tag{4.32}$$
$$T_2(f_2) = q + 8280q^2 + O(q^3) = f_1 - 9216f_2; \tag{4.33}$$

because T_2 preserves the space $S_{28}(\mathrm{SL}_2(\mathbf{Z}))$, we can write $T_2(f_1)$ and $T_2(f_2)$ in terms of f_1 and f_2; we find the coefficients here by linear algebra. Therefore, we have that the matrix of the Hecke operator T_2 with respect to the basis $\{f_1, f_2\}$ we have chosen is

$$\begin{pmatrix} 936 & 18662400 \\ 1 & -9216 \end{pmatrix}.$$

We can now diagonalize this matrix; we find that the eigenvalues are $-4140 \pm 108\sqrt{18209}$ and their respective eigenvectors are

$$f_1 + (-5076 \pm 108\sqrt{18209}) \cdot f_2. \tag{4.34}$$

Because the Hecke operators commute with one another, and because these two eigenspaces for T_2 are already 1-dimensional, these two modular forms are in fact simultaneous eigenforms for *all* the Hecke operators. We note that they have coefficients which are real numbers, verifying the corollary.

We recall an observation from [Koblitz (1993)], Exercise III.5.6, that the discriminant of the field of definition of the simultaneous eigenforms can be large; we can bound the degree of the field extension simply by considering the dimension of the space, but the actual field of definition is more difficult to pin down without actually doing the computation. There are other fields of large discriminant which appear in similar examples. For instance, in the exercise mentioned above, it is shown that the eigenforms for $S_{26}(\mathrm{SL}_2(\mathbf{Z}))$ have Fourier expansions defined over $\mathbf{Q}(\sqrt{144169})$.

We note that there is nothing special about the operator T_2; we could have chosen any Hecke operator T_p to diagonalize, but computing the operator T_2 requires knowledge of the fewest of the Fourier coefficients of f_1 and f_2, so is the best choice for computational purposes. We will pursue this line of thought further in Section 7.5.3.

The choice of a basis $\{f_1, f_2\}$ with the forms having different orders of vanishing at ∞ follows the usage of [Koblitz (1993)], Exercise III.5.6; this makes the linear algebra much easier, and aids computation. The calculation is also aided by starting with a basis of forms which is defined over $\mathbf{Z}$.

In Exercise 1 the reader is invited to work through similar calculations to gain familiarity with this type of argument.

4.2.3 *Integral bases*

In the worked example for $S_{28}(\mathrm{SL}_2(\mathbf{Z}))$ we found two bases of the space of modular forms; we began with a basis with Fourier coefficients in $\mathbf{Z}$, and we found a basis of eigenforms with Fourier coefficients in $\mathbf{Q}(\sqrt{18209})$.

This phenomenon occurs more generally; in Exercise 1 of Chapter 3 we have seen that there is a $\mathbf{Z}$-basis of $M_k(\mathrm{SL}_2(\mathbf{Z}))$ which is also a $\mathbf{C}$-basis, and we have just proved that $M_k(\mathrm{SL}_2(\mathbf{Z}))$ has a basis of eigenforms. We will show that these (normalized) eigenforms have Fourier coefficients which are algebraic integers.

This also aids our computations; working with floating-point numbers leads to ambiguity (it is well-known that floating-point arithmetic is not associative, for instance; see Section 7.5.4.1 for an example), but if we know ahead of time that we are working with integers and we can work them out to a sufficient precision, then we can know them exactly.

Proposition 4.24. *Suppose that a space of modular forms S for $\mathrm{SL}_2(\mathbf{Z})$ has a $\mathbf{Z}$-basis which is also a $\mathbf{C}$-basis. Then it has a basis of eigenforms* *whose Fourier coefficients are real algebraic integers.*

Proof. We consider the action of the Hecke operators T_n on Fourier expansions. From (4.7) we see that if f has integral Fourier coefficients then $T_n f$ does also, so the coefficients of the characteristic polynomials of the T_n are integral, so the eigenvalues of the characteristic polynomial must be algebraic integers. These are real by Corollary 4.23. □

The two-dimensional example that we computed also shows that we can have modular forms whose Fourier expansions are conjugate to one another; we see in equation (4.34) that the two eigenforms are conjugates under the nontrivial automorphism in $\mathrm{Gal}(\mathbf{Q}(\sqrt{18209})/\mathbf{Q})$. This also occurs in general; if f is a modular form in $M_k(\Gamma_0(N), \chi)$ with Fourier expansion $f(q)$ taking values in some finite extension F of $\mathbf{Q}$, and $\sigma \in \mathrm{Gal}(F/\mathbf{Q})$, then

$$\sigma(f)(q) := \sum_{n=0}^{\infty} \sigma(a_n) q^n$$

is the Fourier expansion of a modular form in $M_k(\Gamma_0(N), \sigma(\chi))$, where we define $\sigma(\chi)$ to be the image of χ under σ. This follows by taking the transformation formulae for f and applying σ to both sides of it.

4.3 Oldforms and newforms

We saw in Section 4.2 that we can find bases of spaces of modular forms for all of the Hecke operators T_p which have p not dividing the level, and also that in certain cases there is no basis of eigenforms for all of the T_p. This leads us to ask for which spaces of modular forms we can find bases of simultaneous eigenforms for the Hecke operators.

In this section, we will find a resolution to this problem, by identifying important subspaces which can be spanned by bases of eigenforms. These are called the *newforms*.

From Section 4.1.3, we see that if $N|M$, there are inclusion maps from $M_k(\Gamma_*(N))$ into $\mathcal{M}_k(\Gamma_*(M))$, where $*$ is one of 0, 1 or no symbol. For instance, if p is a prime then we have three maps from $M_k(\Gamma_0(1))$ into $M_k(\Gamma_0(p^2))$, whose effect on Fourier expansions is to send $f(q)$ to one of $f(q)$, $f(q^p)$ or $f(q^{p^2})$.

The following definition will distinguish between those modular forms which are in the image of these maps, and those which are not. One reason why we may wish to do this comes from Exercise 3; if there are "sufficiently many" oldforms, then we no longer have a basis of eigenforms for the whole space of modular forms.

Definition 4.25. Let N be a positive integer, and let d be any divisor of N. We define the inclusion map $\imath_d$ in the following way:

$$\imath_d : S_k(\Gamma_1(N/d)) \times S_k(\Gamma_1(N/d)) \to S_k(\Gamma_1(N)) \tag{4.35}$$

$$(f(z), g(z)) \mapsto f(z) + g(z^d). \tag{4.36}$$

We define the space of *oldforms at level* $\Gamma_1(N)$ to be

$$S_k^{\text{old}}(\Gamma_1(N)) = \sum_{\substack{p|N \\ p \text{ prime}}} \imath_p \left(S_k(\Gamma_1(N/p)) \times S_k(\Gamma_1(N/p))\right),$$

and the space of *newforms at level* $\Gamma_1(N)$ to be the orthogonal complement of the oldforms with respect to the Petersson inner product inside $S_k(\Gamma_1(N))$,

$$S_k^{\text{new}}(\Gamma_1(N)) = (S_k^{\text{old}}(\Gamma_1(N)))^{\perp}.$$

Remark 4.26. In the literature, newforms are sometimes called *primitive forms.* We also say that a modular form is *new* (respectively, *old*) if it is a newform (respectively, an oldform).

We will now consider a few explicit examples of oldforms and newforms, to fix our intuitions.

In certain cases, we know that the space of newforms is in fact all of the space of modular forms; for instance when the level $N = 1$, so there are no prime divisors of N. In that case we have

$$S_k^{\text{new}}(\mathrm{SL}_2(\mathbf{Z})) = S_k(\mathrm{SL}_2(\mathbf{Z})).$$

We notice that when the level is 1 the restrictions given in the theorems concerning the Petersson inner product and the Hecke operators do not apply, as there are no primes dividing the level. Therefore we can already deduce that there exists a basis of simultaneous eigenforms for all of the Hecke operators in level 1.

Another well-known example occurs when we have modular forms of weight 2 and level $\Gamma_0(p)$, where p is a prime. We have shown in that there are no nonzero modular forms for $\mathrm{SL}_2(\mathbf{Z})$ with weight 2 (see Theorem 3.5), so there can be no oldforms.

Conversely, there are examples of spaces of modular forms where there are only oldforms, and no newforms. For instance, consider $S_{12}(\Gamma_0(2))$. In Exercise 14, we will see (by dimensional considerations) that

$$S_{12}^{\text{old}}(\Gamma_0(2)) = S_{12}(\Gamma_0(2)) \text{ and } S_{12}^{\text{new}}(\Gamma_0(2)) = 0.$$

It can be seen that, for a fixed congruence subgroup Γ, there are only finitely many spaces $S_k(\Gamma)$ which are completely composed of oldforms, because the rate of growth of the modular forms for Γ is faster than the rate of growth of modular forms for congruence subgroups containing Γ.

We note that the zeroes of a modular form may change depending on whether it is viewed as either an oldform or a newform. For instance, the residue formula tells us that the Δ function has a unique zero at ∞, but if we view it as a modular form for $\Gamma_0(p)$ for a prime p, then because it is a cuspform it vanishes at the cusps 0 and ∞ of $\Gamma_0(p)$. This follows because under the action of $\mathrm{SL}_2(\mathbf{Z})$ the set $\mathbf{Q} \cup \{\infty\}$ forms one equivalence class, but under the action of $\Gamma_0(p)$ it forms two, so the zero "splits" in level p.

This reminds us that a modular form is always (implicitly) viewed as having a certain level; usually it is implicit from context, but sometimes it has to be stated explicitly to avoid confusion.

One very important property of the spaces of oldforms and newforms is that they are both stable under the action of the Hecke operators; if f is an oldform, then $T_n f$ is also an oldform, and similarly if f is a newform then $T_n f$ is also a newform.

We prove this in two parts. The easier part is to show that the oldforms are stable under the action of the Hecke operators.

Proposition 4.27. *Let $f \in S_k^{\text{old}}(\Gamma_1(N))$ be a modular form and let T_n be a Hecke operator, with $(n, N) = 1$. Then $T_n f \in S_k^{\text{old}}(\Gamma_1(N))$.*

Proof.

By definition, if f is an oldform then we can write it as a sum of forms with level dividing N, as follows:

$$f(q) = \sum_{d|N} f_d(q^d), \text{ with } f_d \in S_k(\Gamma_1(N/d)).$$

We define $g_d(q) := f_d(q^d)$ to make our notation easier to read. Because n and Nd have no common factor, we have that

$$(T_n f)(q) = \sum_{d|N} (T_n g_d)(q) \tag{4.37}$$

$$= \sum_{d|N} (T_n f_d)(q^d). \tag{4.38}$$

We know that $T_n f_d(q)$ is an element of $S_k(\Gamma_1(N/d))$ from the definition of the Hecke operators, and the map whose action on Fourier expansions is $q \mapsto q^d$ maps modular forms from $S_k(\Gamma_1(N/d))$ into $M_k(\Gamma_1(N))$, and specifically into $S_k^{\text{old}}(\Gamma_1(N))$, so we are done. □

Now we will prove the corresponding result that the new subspace is preserved by the Hecke operators.

Proposition 4.28. *Let $f \in M_k^{\text{new}}(\Gamma_1(N))$ be a modular form and let T_n be a Hecke operator, with $(n, N) = 1$. Then $T_n f \in M_k^{\text{new}}(\Gamma_1(N))$.*

Proof. The defining characteristic of a newform f is that it is orthogonal to the space of oldforms with respect to the Petersson scalar product; in other words, that

$$\langle f, g\rangle = 0 \text{ for all } g \in M_k^{\text{old}}(\Gamma_1(N)).$$

We will now use the self-adjointness property of the Hecke operators:

$$\langle T_n f, g\rangle = \chi(n) \cdot \langle f, T_n g\rangle,$$

and by Proposition 4.27 we see that oldforms are preserved by T_n, so we have that $\langle f, T_n g\rangle = 0$, as it is the inner product of an oldform and a newform. This means that $T_n f$ is a newform also, which is what we needed to show. □

We also record a useful result that we will need later, which gives us another interesting way of finding oldforms.

Theorem 4.29. *Let N be a positive integer and let χ be a Dirichlet character of conductor N. Let M be another positive integer and let ε be a Dirichlet character of conductor M. Let $f \in M_k(\Gamma_0(N), \chi)$ be a modular form with Fourier expansion $f(q) = \sum_{n=0}^{\infty} a_n q^n$. Then there is a modular form $f_\varepsilon \in M_k(\Gamma_0(NM^2), \chi\varepsilon^2)$ with*

$$f_\varepsilon(q) = \sum_{n=0}^{\infty} \varepsilon(n) a_n q^n.$$

We say that f_ε is f twisted by *ε.*

Proof. This can be found in [Shimura (1994)] as Proposition 3.64 and in [Koblitz (1993)] as Proposition III.17.b. We follow the account given in Shimura. Let ζ be an Mth root of unity and let j be an integer. We define a matrix $\alpha_j := \left(\begin{smallmatrix} 1 & j/M \\ 0 & 1 \end{smallmatrix}\right)$, and a Gauss sum $W(\chi)$ to be

$$W(\chi) := \sum_{c=0}^{r-1} \chi(c) e^{(2\pi i c)/r}.$$

We see by explicitly computing the Fourier series that

$$(f|[\alpha_j]_k)(q) = \sum_{n=0}^{\infty} a_n e^{2\pi i n(z+j/r)} = \sum_{n=0}^{\infty} \zeta^{jn} a_n q^n.$$

In fact, using Exercise 19, we see that

$$W(\overline{\chi}) \cdot f_\varepsilon(q) = \sum_{i=1}^{N} \overline{\chi}(i) \cdot f|[\alpha_i]_k.$$

We now use Proposition 2.20 to show that each of the $f|[\alpha_j]_k$ is a modular form for the congruence subgroup $\Gamma_1(NM^2)$. Finally, we check that f_ε actually is a modular form for $\Gamma_0(NM^2)$ with character $\chi\varepsilon^2$, by the following explicit means, again following the argument in [Shimura (1994)].

We take $\gamma := \left(\begin{smallmatrix} a & b \\ cN & d \end{smallmatrix}\right) \in \Gamma_0(NM^2)$, and define

$$\delta := \begin{pmatrix} a + cjN/M & \beta \\ cN & d - cd^2jN/M \end{pmatrix},$$

where

$$\beta := b + dj(1 - ad)/M - cd^2u^2N/M^2.$$

We have that the bottom right corner of δ is congruent to d modulo N (because c is congruent to 0 modulo M^2), and that δ has integral entries (because ad is congruent to 1 modulo NM^2), so therefore

$$\begin{pmatrix} 1 & j/M \\ 0 & 1 \end{pmatrix} \cdot \begin{pmatrix} a & b \\ cN & d \end{pmatrix} = \delta \cdot \begin{pmatrix} 1 & d^2j/M \\ 0 & 1 \end{pmatrix}.$$

In fact, we could *define* the matrix δ by this equation.

This means that we have (if we set $l = d^2j$) that $f|[\alpha_j\gamma]_k = (f|[\delta]_k)|[\alpha_l]_k = \chi(d)f|[\alpha_j]_k$, because f is a modular form for $\Gamma_0(N)$, so we have that

$$g|[\gamma]_k = W(\overline{\chi})^{-1}\chi(d)\varepsilon(d^2)\sum_l f|[\alpha_l]_k = \chi(d)\varepsilon(d)^2 g,$$

so g transforms correctly under the action of an arbitrary $\gamma \in \Gamma_0(NM^2)$, so therefore it is a modular form for this subgroup with the correct character, which is what we wanted to show. □

4.3.1 *Multiplicity one for newforms*

We have seen in Exercise 3 that there are spaces of modular forms for which there is no basis of simultaneous eigenforms for all of the Hecke operators; the presence of a substantial number of oldforms made it impossible to find a basis of eigenforms for a T_p with $p|N$. In this subsection, we will review results which show that if we consider only a space of newforms then we will always be able to find a basis of eigenforms for *all* the Hecke operators.

This uses the work of Atkin and Lehner from [Atkin and Lehner (1970)], as extended by Miyake in [Miyake (1971)] and Li in [Li (1975)]; general references for this subject are [Lang (1976)], Chapter VIII and [Diamond and Shurman (2005)], Section 5.8.3. The reader should be aware that the "character" mentioned in Theorem VIII.3.3 of Lang is the map which maps an eigenfunction f and a Hecke operator T_n to the eigenvalue of T_n acting on f.

We quote the following theorem from [Lang (1976)], where it appears as Theorem VIII.3.1.

Theorem 4.30. *Let $f \in S_k(\Gamma_1(N))$ with Fourier expansion $f(q) = \sum_{n=0}^{\infty} a_n q^n$. Suppose that there exists an integer $M \geq 1$ such that $a_n = 0$ if $(MN, n) = 1$. Then there are modular forms $g_d \in S_k(\Gamma_1(N/d))$ with $d|N$ and Fourier expansions $g_d(q)$ such that*

$$f(q) = \sum_{d|N} g_d(q^p).$$

Proof. A proof of this theorem is given in Section VIII.4 of [Lang (1976)]. It involves a careful analysis of the action of the V_p operator on the Fourier expansions of modular forms. □

One of the first ingredients for the multiplicity one theorem is a result which tells us that certain eigenforms are oldforms.

Theorem 4.31. *Let N be an integer and let $\chi : (\mathbf{Z}/N\mathbf{Z})^\times \to \mathbf{C}^\times$ be a Dirichlet character. Let $f \in M_k(\Gamma_0(N), \chi)$ be a nonzero eigenform with Fourier expansion $\sum_{n=0}^{\infty} a_n q^n$ for the Hecke operators T_m with $(m, N) = 1$. If $a_1 = 0$ then we have that $a_n = 0$ for $(n, N) = 1$, and hence that f is an oldform.*

Proof. Let $p \nmid N$ be a prime number, and let λ be the eigenvalue of T_p acting on f. By the definition, if n is a positive integer then

$$a_{np} + \chi(p)p^{k-1}a_{n/p} = \lambda \cdot a_n,$$

where we take $a_{n/p} = 0$ if n is not divisible by p. We see that $a_p = 0$, by substituting in $n = 1$ and using the fact that $a_1 = 0$, and then by induction we see that every $a_{p^r} = 0$.

To obtain the result for general n, we recall that T_n is defined as the product of the T_{p^r} such that $p^r || n$. Therefore, if we have that $T_{p^r} = 0$, we see that T_n must be zero also, which is what we wanted to show.

Finally, we apply Theorem 4.30 to show that f is an oldform. □

It should be noted that just because a modular form for $\Gamma_1(N)$ is an eigenform for the Hecke operators T_n with $(n, N) = 1$, this does not mean that it is a newform. For instance, consider $\Delta(q)$ as an element of $S_{12}(\Gamma_1(2))$. This is an eigenform for T_n for every odd n, but it is not a newform.

We can now prove the following important result about the multiplicities of eigenforms, which is Theorem 3.3 of Chapter VIII of [Lang (1976)].

Theorem 4.32 (Multiplicity one). *Let N be a positive integer, and let f and g be two eigenforms for the Hecke operators T_n with $(n, N) = 1$ with the same eigenvalues. If f is a nonzero newform then there exists a constant $\lambda \in \mathbf{C}$ such that $f = \lambda \cdot g$. If in addition g is an oldform, then $g = 0$.*

Proof. We will use the notation of [Lang (1976)] here.

By the definitions we have given for newforms and oldforms, any modular form g can be written as a sum of an oldform and a newform, so $g = g_o + g_n$, where g_o is an oldform and g_n is a newform. As g_o and g_n have the same character, it will be enough to prove the theorem in the special cases when g is a newform and when g is an oldform.

If g is a nonzero newform, then without loss of generality we may assume that f and g are both normalized. This means that $f - g$ is an eigenform and a newform with Fourier coefficient $a_1 = 0$, and therefore by Theorem 4.31 we see that $f - g = 0$, so $f = g$ as required.

If g is a nonzero oldform then because we can find a basis of eigenforms for the T_p with $p \nmid N$ we can write it as a sum of the form

$$g = \sum_{i=1}^{n} a_i g_i, \ a_i \in \mathbf{C},$$

where the g_i are old eigenforms for T_p with $p \nmid N$ which have the same character as g. We assume that g_i appears with non-zero coefficient in g (if there is no such g_i then g is zero, which is a contradiction). We can write g_i as the image of a modular form h under the degeneracy maps from lower levels (because it is an oldform).

We can use Theorem 4.31 on the modular form h (viewed as a newform) to show that it has a Fourier expansion with $a_1 \neq 0$. This means that there is a constant C such that $f - Ch$ has Fourier coefficient $b_1 = 0$, because f is a newform and therefore has $a_1 \neq 0$, and $f - Ch$ is an eigenfunction for all of the T_p with $p \nmid N$, because both f and h are eigenfunctions with the same eigenvalues.

By Theorem 4.31, we see that this means that $f - Ch$ must be an oldform, and therefore f must be an oldform, but f is also a newform so $f = 0$. This is a contradiction to our assumption that g was nonzero, so we are done. □

Remark 4.33. Theorem 4.32 is called a *multiplicity one theorem* because it shows that each eigenspace in the space of newforms has dimension one. This is in clear contrast to the space of oldforms, where an eigenspace can appear with an arbitrarily high multiplicity. For instance, if we consider $S_2(\Gamma_0(2^n \cdot 11))$, then there is a space of oldforms $\{f(q^{2^i})\}_{i=0,\ldots,n-1}$ which has multiplicity n, where $f \in S_2(\Gamma_0(11))$ is the unique normalized eigenform. The failure of multiplicity one can lead to unwanted effects, such as the failure to have a basis of eigenforms for all of the T_p; see Exercise 3.

Remark 4.34. We also note that this theorem relies heavily on the fact that we are working in characteristic 0. In Chapter 6, we will see that the naïve generalization of multiplicity one to characteristic p can fail to hold in certain cases.

The naïve generalization of multiplicity one to the mod p setting already runs into the problem that we have modular forms which are congruent to each other modulo certain primes; for instance, there are two eigenforms f_1 and f_2 which span $S_2(\Gamma_0(37))$, and it can be shown that they have Fourier expansions which are congruent modulo 2.

We now prove that, if we have a space of newforms, then we can find a basis of simultaneous eigenforms for all of the Hecke operators T_n. A good although brief reference for this is given in [Milne (1997a)], Chapter 26.

Theorem 4.35 (Basis of simultaneous eigenforms for all the T_n). *Let N be a positive integer and let p be an arbitrary prime number. The space $S_k^{\text{new}}(\Gamma_1(N))$ has a basis of simultaneous eigenforms for all T_p, and therefore for every Hecke operator T_n.*

Proof.

We can use the spectral theorem to write $S_k^{\text{new}}(\Gamma_1(N))$ as a finite sum of eigenspaces B_i for the T_p with $p \nmid N$, which gives us a basis B of eigenforms for the T_p with $p \nmid N$. We will show that this basis is also a basis of eigenforms for *all* the T_p.

Because we are dealing with a space of newforms, we can use Theorem 4.32 to show that each of the B_i has dimension 1; we are using the multiplicity one property here. This is the step which could break down if we had oldforms present.

We now see that the Hecke operators T_p with $p|N$ preserve each of the spaces B_i, because the Hecke operators commute with one another. This means therefore that the basis elements are also eigenforms for the T_p with $p|N$, and therefore for all of the Hecke operators T_n. □

4.4 Exercises

(1) Choose some spaces of modular forms with small dimension, such as $S_{30}(\mathrm{SL}_2(\mathbf{Z}))$ or $S_{14}(\Gamma_0(2))$, and compute a basis of eigenforms in the following way (you may wish to use a computer for this):

   (a) Find a basis for the space S of modular forms. You may want to make sure that the elements of this basis have coefficients in $\mathbf{Z}$, or different orders of vanishing at ∞, or some other useful property.
   (b) Compute the first few Fourier coefficients of each element of this basis. Use the results of this chapter to work out how many coefficients you need to compute.
   (c) Choose a Hecke operator T and compute the action of this Hecke operator on the Fourier coefficients of each element f of your basis of S. Write $T(f)$ in terms of the elements of S.
   (d) Write down the matrix of T with respect to the chosen basis of S, and find its eigenvalues and eigenvectors.
   (e) Check that the eigenvectors that you found in the last step are actually eigenvectors for T.

(2) Let f be a normalized eigenform for all of the Hecke operators T_p with Fourier expansion $f(q) = \sum_{n=0}^{\infty} a_n q^n$. Show that $a_1 \neq 0$. Show that if f is a Hecke eigenform for all of the T_p except one then a_1 can equal 0.

(3) In this exercise, we will show that there exist spaces of modular forms which do *not* have a basis of simultaneous eigenforms for all of the

Hecke operators.

(a) Let $f_0(z) \in S_2(\Gamma_0(19))$ be the unique normalized eigenform. Show that this space of modular forms is 1-dimensional, and explain briefly why $f_0(z)$ is an eigenform for all of the Hecke operators.
(b) Verify that the coefficient of q^3 is -2.
(c) Define $f_1(z) = f_0(3z)$, $f_2(z) = f_0(9z)$, $f_3(z) = f(27z)$. Check that the f_i are linearly independent in $S_2(\Gamma_0(513))$.
(d) Define $V := \{f_0, f_1, f_2, f_3\}$. Show that this space is stable under the action of the Hecke operators, paying particular attention to T_3 and T_{19}. Why is f_0, when considered as a modular form of level 513, not an eigenform for T_3?
(e) Compute the matrix of T_3 acting on the space V. Show that there are exactly three normalized eigenforms for T_3 within V, and hence that it does not have a basis of eigenforms.
(f) Find other examples of this phenomenon, in other levels and weights and for other Hecke operators T_p. What properties of the matrix are we using? Can you prove a general theorem about these?

(4) In this exercise we will go over the details of the proof of Theorem 4.2, part (2). Let m and n be positive integers.

(a) Assume that m and n are coprime. Show that one can indeed write each of the matrices comprising T_{mn} as a product of a matrix from T_m and a matrix from T_n, and hence show that that $T_{mn} = T_m \cdot T_n$.
(b) Go through the details of the induction proof for (4.4).
(c) Finally, check that we can combine these to prove the result for general m and n.

(5) Find spaces of modular forms $S_k(\Gamma_0(N))$ of dimension at least 2 such that the Hecke algebra over $\mathbf{Z}$ generated by the $\{T_p\}$ up to the Sturm bound is *not* the same as the Hecke algebra over $\mathbf{Z}$ generated by the $\{T_n\}$ up to the Sturm bound, in the following way:

(a) Generate a list of such spaces with dimension 2, say.
(b) Compute the Sturm bound for each of these.
(c) Compute both the full Hecke algebra and the anaemic Hecke algebra over $\mathbf{Z}$ for each space, and compare the two.

(6) Compute the Hecke algebra over $\mathbf{Q}$ generated by the T_n acting on $S_2(\Gamma_0(37))$.

(a) Compute the Sturm bound and work out how many Hecke operators will be needed, and work out the dimension of this space of modular forms.

(b) Using MAGMA or otherwise, compute some Hecke operators and hence find the Hecke algebra $\mathbf{T}_{\mathbf{Z}}(S_2(\Gamma_0(37)))$.

(7) If $f \in M_k(\Gamma_0(N), \chi)$ and $g \in M_k(\Gamma_0(N), \mu)$ have different characters then f and g are orthogonal with respect to the Petersson scalar product.

(a) Verify that $F := \langle f, g \rangle = \langle f|[\alpha]_k, g|[\alpha]_k \rangle$ for $\alpha \in \mathrm{GL}_2^+(\mathbf{Q})$.

(b) Choose an α so that $|[\alpha]_k$ is a useful diamond operator and rewrite F so that the action of the diamond operator is outside the Petersson inner product.

(c) Use this to show that $\langle f, g \rangle$ has to be zero.

(8) Check the computations in (4.11); in particular, show that the second integral does evaluate to $2/\sqrt{3}$.

(9) Let p be a prime number. Show that $E_2^*(q) := E_2(q) - pE_2(q^p)$ is the Fourier expansion of a modular form of weight 2 for $\Gamma_0(p)$.

(a) Show that E_2^* transforms correctly under the action of $\Gamma_0(p)$.

(b) Show that E_2^* has no poles at the cusps of $\Gamma_0(p)$.

(c) Verify that E_2^* has no poles on $\mathcal{H}$.

(10) Let $k \geq 4$ be an even integer, let $p \geq 2$ be a prime and let $E_2^*(q)$ be the modular form defined in Question 9. Show that E_2^* is *not* a cusp form, despite being having no constant term, by using the Ramanujan-Petersson conjecture or otherwise. Explain how this is possible.

(11) In this exercise we will find modular eigenforms f and g such that $f \cdot g$ is also a modular eigenform. This subject is discussed in [Emmons (2005)].

Write down relations between the coefficients of f, g and $f \cdot g$.

Here are some guidelines for finding these:

(a) It will be useful to have f, g and $f \cdot g$ contained within spaces of modular forms with small dimension.

(b) Consider the coefficients of f and g. Can you put a constraint on either f or g? In particular, can both f and g have nonzero constant terms?

(12) Show that there are no cusp forms of weight 1 and level $N < 23$.

(a) Show that the product of two cusp forms is a cusp form.

(b) Using Question 2, consider whether the product of two cusp forms is an eigenform, and recall that the Hecke operators T_p preserve the space of cusp forms.

(c) Use this to bound the size of the space of cusp forms of weight 2 and level N from below.

(d) Now use the dimension formulae given to prove the result, except for level 22.

(e) To deal with level 22, find a basis for $S_2(\Gamma_0(22))$ of the form $f(q) = q + O(q^2)$, $g(q) = q^2 + O(q^3)$, and show that $g(q)$ is not the square of a Fourier expansion in q with rational Fourier coefficients.

(13) Show that the meromorphic functions on $\mathcal{H}$ form a field.

(14) In this exercise we will show that $S_{12}(\Gamma_0(2)) = S_{12}^{\text{old}}(\Gamma_0(2))$ and therefore that $S_{12}^{\text{new}}(\Gamma_0(2)) = 0$.

(a) Using the dimension formulae, or otherwise, show that $\dim S_{12}(\Gamma_0(2)) = 2$.

(b) Find two different ways to embed $S_{12}(\mathrm{SL}_2(\mathbf{Z}))$ into $S_{12}(\Gamma_0(2))$.

(c) Show that these two different modular forms are linearly independent.

(15) In a similar way to Exercise 14, show that $S_8(\Gamma_1(4)) = S_8^{\text{old}}(\Gamma_1(4))$.

(16) Let Γ be a congruence subgroup and let k be an odd positive integer. Show that if $-I \in \Gamma$ then $M_k(\Gamma) = S_k(\Gamma) = 0$. For which of the standard congruence subgroups $\Gamma_0(N)$, $\Gamma_1(N)$, and $\Gamma(N)$ is $M_k(\Gamma)$ identically 0?

(17) Let $\alpha = \left(\begin{smallmatrix} a & b \\ c & d \end{smallmatrix}\right) \in \mathrm{GL}_2(\mathbf{R})$. Show that $d\mu(\alpha z) = d\mu(z)$ for every $z \in \mathcal{H}$, by proving that

$$\frac{d\alpha z}{dz} = \frac{\det \alpha}{(cz+d)^2}$$

and then using the chain rule, or otherwise.

(18) Recall that if ε is a Dirichlet character, and f is a modular form with Fourier expansion $f(q) = \sum_{n \in \mathbf{N}} a_n q^n$ then the modular form f_ε from Theorem 4.29 is defined to have Fourier expansion

$$f_\varepsilon(q) = \sum_{n=0}^{\infty} \varepsilon(n) a_n q^n.$$

Show that we can write the modular form f_ε in terms of the $f \| [\alpha_i]_k$, by considering how to write the character ε in terms of the powers of the root of unity ζ, or otherwise.

(19) In this exercise we prove some useful subsidiary results on Gauss sums to help us prove Theorem 4.29.

(a) Show that, if $j \in \mathbf{Z}$ and χ is a Dirichlet character modulo N, that

$$\sum_{i=0}^{N-1} \chi(i)\zeta^{ij} = \overline{\chi}(j)W(\chi).$$

(b) For any χ, we have that $\overline{W(\chi)} = \chi(-1)W(\overline{\chi})$.

(20) Let $n \in \mathbf{N}$. Prove that if $f \in M_k(\Gamma_0(N), \chi)$ is an eigenform for the Hecke operator T_n and the constant term in the Fourier expansion of f is nonzero, then the eigenvalue for T_n is given by

$$c_m := \sum_{1 \leq d | m} \chi(d)d^{k-1}.$$

You may wish to consider (4.7).

(21) Let τ be the nontrivial Dirichlet character of conductor 4. Show that

$$\begin{aligned} M_k(\Gamma_1(4)) &= M_k(\Gamma_0(4)), \text{ if } k \text{ is even}, \\ &= M_k(\Gamma_0(4), \tau), \text{ if } k \text{ is odd}. \end{aligned}$$

(22) Let $f = \Delta \cdot E_4 \in S_{16}(\mathrm{SL}_2(\mathbf{Z}))$. Find eigenvectors for the action of the T_2 operator on the subspace of $S_{16}(\Gamma_0(2))$ generated by the image of f under the two inclusion maps. Choose other eigenvectors f at level N and a prime $p \nmid N$ and compute a basis of eigenvectors spanning the space $\{f(q), f(q^p)\}$.

(23) Can one extend Theorem 4.35 to spaces of modular forms $S_k(\Gamma_0(N))$, where N is cubefree? For instance, let k be an even integer and let p be a prime, and consider spaces of modular forms of the form $S_k(\Gamma_0(p^2))$.

Chapter 5

Applications of modular forms

> There is no branch of mathematics, however abstract, which may not some day be applied to phenomena of the real world. — Nikolai Lobachevsky.

In the previous chapter we laid some foundations for the study of modular forms for the sake of their arithmetic. In this chapter we will consider some classical and modern applications of modular forms and modular functions: these will include

1. finding the number of representations of an integer by a quadratic form,
2. approximating π,
3. proving Picard's Little Theorem,
4. finding identities for sums and products of arithmetic functions, and
5. bounding the number of points on an elliptic curve over a finite field.

Many of these results depend strongly on the finite-dimensional nature of a given space of modular forms $M_k(\Gamma)$ for a fixed weight k and congruence subgroup Γ, which we proved in Chapter 3, using the complex-analytic properties of modular forms.

We will also consider certain special types of modular forms; the lacunary modular forms, which have large numbers of Fourier coefficients which are zero, and forms with complex multiplication. We also consider modular forms of half-integral weight, which appear naturally when one considers the question of representing integers by quadratic forms in an *odd* number of variables. Where suitable, we will prove our results; in other cases, we include references to the literature in the text so that the reader will know where to look to learn more.

We will also introduce the modular *functions*; these are the fraction field of the modular forms. Because the space of modular functions of any given weight is infinite-dimensional, we have to be careful with our analysis of modular functions; we must be careful when trying to show that a given modular function is zero. We will define an important modular function, the j-invariant, and use this to explain a numerical coincidence similar to those noted by Ramanujan.

We will also briefly consider the celebrated proof of Fermat's Last Theorem by Andrew Wiles; this relies crucially on the arithmetic of modular forms, and the link between modular forms and elliptic curves.

Some references for this chapter are [Koblitz (1993)], Chapter IV, for the discussion of modular forms of half-integral weight, the introductory chapter of [Cornell *et al.* (1997)], for the discussion of Fermat's Last Theorem, and [Miyake (2006)], for the discussion of the modularity of theta functions.

5.1 Modular functions

In the definition of a modular form f, we have three conditions; that it transforms correctly under the action of a congruence subgroup Γ, that it is holomorphic at ∞ and any other cusps of Γ, and that it is holomorphic on the half-plane $\mathcal{H}$. It is natural to wonder what would happen if we weakened these conditions; we would like to retain the transformation properties, but what happens if we do not require f to be holomorphic at ∞, for instance?

It follows from the definition that if f and g are modular forms of weight k, with $g \neq 0$, then f/g is weakly modular of weight 0. We see that f/g is a modular form if for every zero of g at a point z_0 which has multiplicity n there is a zero of f at z_0 which has multiplicity at least n; we used this in Chapter 3 to find dimension formulae for modular forms for $\mathrm{SL}_2(\mathbf{Z})$, by dividing cusp forms of level 1 by the Δ function, which has a unique zero at ∞.

We now present the definition of a modular function.

Definition 5.1 (Modular Functions). *Let k be an integer, let Γ be a congruence subgroup and let $f : \mathcal{H} \to \mathbf{C}$ be a meromorphic function which is weakly modular of weight k for Γ. We say that f is a* modular function *of weight k for Γ.*

Remark 5.2. The notation for modular functions varies by author. Some allow modular functions to be of any weight, while others always assume that a modular function is a modular function of weight 0. We will allow modular functions to have any weight.

Before we begin our study of modular functions of weight 0, we note here that if f is a nonzero modular form of weight k, then $1/f$ is a modular function of weight $-k$; the weakly modular property is clear, and the meromorphicity of $1/f$ follows by considering the holomorphicity and transformation properties of f.

The classic (and canonical) example of a modular function of weight 0 which is *not* a modular form is the *modular j-invariant*, which is defined as

$$j(z) := \frac{E_4(z)^3}{\Delta(z)} = q^{-1} + 744 + 196884q + \cdots$$

It is also sometimes known (especially in older work) as "Klein's j-invariant", or "Klein's absolute invariant".

We see that the j-invariant is not a modular *form* because it has a simple pole at ∞, corresponding to the unique zero of the Δ function. We note that this could also be deduced from the dimension formulae; the space $M_0(\mathrm{SL}_2(\mathbf{Z})) = \mathbf{C}$, so as $j(z)$ is not constant it cannot be a modular form. On the other hand, we can see that it is a modular *function* of weight 0 because it is a quotient of two modular forms of the same weight, so transforms correctly under the action of $\mathrm{SL}_2(\mathbf{Z})$ and is meromorphic on $\mathcal{H}$ and at ∞.

Also, we see that any power of the j-invariant is also a modular function of weight 0, by considering its transformation formula. This means that the ring of modular functions of weight 0 is an infinite-dimensional complex vector space, which is a very different situation to the world of modular forms, where the space of modular forms of weight k is finite-dimensional for any integer weight k. We will discuss this further in Chapter 7.

We will now show that every modular function of weight 0 for the full modular group $\mathrm{SL}_2(\mathbf{Z})$ is a rational function of j.

Theorem 5.3. *The modular functions of weight 0 for* $\mathrm{SL}_2(\mathbf{Z})$ *are exactly the rational functions of* j.

Proof. If f is a rational function of j then it is a quotient of two polynomials in j. Each of these polynomials is a modular function of weight 0 (note that the constant term is also a modular function of weight 0), and from the definitions we see that the quotient of these two modular functions of weight 0 is also a modular function.

Conversely, let f be a modular function of weight 0 for $\mathrm{SL}_2(\mathbf{Z})$. By considering the residue formula (3.1), which is valid for modular *functions* as well as for modular forms, we see that f has only finitely many poles in a fundamental domain F for $\mathrm{SL}_2(\mathbf{Z})$. Let $\{p_i\}$ be the poles of f in F counted with multiplicity, and define

$$g(z) := f(z) \cdot \prod_i (j(z) - j(p_i)).$$

This function has no poles on $\mathcal{H}$, because each pole of f is cancelled out by a zero in the product, and j has a unique pole at ∞. It is also a modular function, because $j(p_i)$ is a constant, and therefore a modular form of weight 0. If we can show that g is a rational function of j, then we will have shown that f is a rational function of j too. So without loss of generality we can assume that f has no poles in $\mathcal{H}$.

We will now cancel the pole at ∞ by multiplying by a sufficiently large power of Δ, which we recall has a simple zero at ∞. So there exists an integer k such that $\Delta^k \cdot f(z) \in M_{12k}(\mathrm{SL}_2(\mathbf{Z}))$.

From Proposition 3.6, we see that we can write $f(z)$ as a linear combination of classical modular functions of the form $E_4^a \cdot E_6^b/\Delta^k$, where $4a + 6b = 12k$, so now we need to check that each of these modular functions is a rational function of the j-invariant. We have that $3|a$ and $2|b$, because 12 divides $4a + 6b$, so in fact we only have to check that E_4^3/Δ and E_6^2/Δ are rational functions of j, because we can write a$E_4^a \cdot E_6^b/\Delta^k$ as a polynomial in these two, simpler, modular functions. This follows because $E_4^3/\Delta = j$ and $E_6^2/\Delta = j - 1728$, so we are done. □

Finally, we will quote the following theorem on the field of modular functions of weight 0 for $\Gamma_0(N)$ from [Milne (1997b)] (Theorem 6.1):

Theorem 5.4. *Let N be a positive integer and let f be a modular function of weight 0 for $\Gamma_0(N)$. We can write f as a rational function in $j(z)$ and $j(Nz)$.*

Proof. [Proof Sketch] There are two parts to the proof given in our reference; first, one must show that both $j(q)$ and $j(q^N)$ are modular functions for $\Gamma_0(N)$, and secondly, one must show that they generate the field of modular functions.

The first of these can be proved in one (fairly long) line of working: following Milne's proof, we see that if we let $\gamma = \left(\begin{smallmatrix} a & b \\ Nc & d \end{smallmatrix}\right) \in \Gamma_0(N)$, then

$$j(N\gamma z) = j\left(\frac{Naz + Nb}{Ncz + d}\right) = j\left(\frac{a(Nz) + bN}{c(Nz) + d}\right) = j(Nz),$$

because $\left(\begin{smallmatrix} a & bN \\ c & d \end{smallmatrix}\right) \in \mathrm{SL}_2(\mathbf{Z})$, and j is a modular function of weight 0 for $\mathrm{SL}_2(\mathbf{Z})$.

To show that these generate all of the modular functions, one uses the fact that $\Gamma_0(N)$ has finite index inside $\mathrm{SL}_2(\mathbf{Z})$ and considers the action of coset representatives of $\Gamma_0(N)$ on $j(z)$. We refer to Chapter 6 of [Milne (1997b)] for the details. □

We also note that we can extend the definition of the Hecke operators to modular functions. From Section 2.3 of [Ono (2004)], we know that the T_n operator acts in the following way on modular functions; if $f(q) = \sum_{n=m}^{\infty} a_n q^n$ is the Fourier expansion of a modular function of weight 0, then the normalized Hecke operator T_p has the following effect on Fourier expansions:

$$T_p f(q) = \sum_{n=pm}^{\infty} (pa_{np} + a_{n/p}) q^n,$$

where as usual $a_{n/p} = 0$ if n/p is not an integer.

We see that this is a modular function of weight 0 by generalizing the definition of Hecke operators that we have given for modular forms to modular functions.

In particular, if f is a modular function for Γ of weight k with Fourier expansion $\sum_{n=m}^{\infty} a_n q^n$, then $T_p f$ is also a modular function for Γ of weight k. We also have the operators U_p and V_p:

$$U_p(f)(q) = \sum_{n=pm}^{\infty} a_{pn} q^n \text{ and } V_p(f)(q) = \sum_{n=pm}^{\infty} a_n q^{pn}.$$

We note that, as for modular forms with trivial character, if $f \in M_k(\Gamma_0(N), \chi)$ then $U_p f$ and $V_p f$ are both in $M_k(\Gamma_0(Np), \chi)$, where the second χ is understood to be a Dirichlet character modulo Np.

By considering the action of Hecke operators on modular functions of weight 0, congruences involving the Fourier expansion of the j-function have been proved by [Coogan (2005)] amongst others.

The following result on modular functions with a pole only at ∞ will prove useful in a later chapter.

Proposition 5.5. *Let g be a modular function for $\mathrm{SL}_2(\mathbf{Z})$ of weight 0, with no poles on $\mathcal{H}$ and a pole of order r at ∞. Then we can write g as a polynomial in the j-invariant of degree r.*

Proof. We will prove this result by induction. If $r = 0$ then g is a modular *form*, so it is a constant. We now assume that we have proved the result for all $m < r$, and let g be a modular function with a pole of order r at ∞ and no poles on $\mathcal{H}$. We now consider the function $g - \alpha j^r$, where α is the Fourier coefficient of q^{-r} in the Fourier expansion of g. This is a modular function with no poles on $\mathcal{H}$ and a pole of order strictly less than r at ∞, so it is covered by the induction hypothesis and therefore we are done. □

We note that in particular $T_p j$ is a polynomial of degree p in j.

5.2 η-products and η-quotients

So far, we have considered Fourier expansions as infinite *sums*. There are other ways of writing the Fourier expansions of modular forms as infinite *products*; we can obtain some interesting equivalences between sums and products in this way.

In order to do this, we define the *Dedekind η function*[1], also known as the η function.

$$\eta(z) = q^{1/24} \prod_{n=1}^{\infty} (1 - q^n), \text{where } q := \exp(2\pi i z). \tag{5.1}$$

We note that we will be considering η as either a function of z or a function of q, depending on context.

This is a function from $\mathcal{H}$ to $\mathbf{C}$ which is holomorphic on $\mathcal{H}$ and has no zeroes on $\mathcal{H}$. It is not a modular form for any $\Gamma_1(N)$, but we will see that we can construct modular forms using the η-product.

Definition 5.6. Let N be a positive integer, let $\{r_\delta\}$ be a set of integers, and let f be a meromorphic function from $\mathcal{H}$ to $\mathbf{C}$ of the form

$$f(z) = \prod_{0<\delta|N} \eta\left(q^\delta\right)^{r_\delta}.$$

We call f an *η-quotient*, and if all of the r_δ are non-negative we say that f is an *η-product.*

There are theorems that tell us that certain η-products and η-quotients are modular forms; for instance, we quote the following result of Ligozat without proof:

[1] This is called the *Dedekind* η product to distinguish it from other η-products; for instance, there is a Jacobi η-product, defined in Section 5.6 of [Armitage and Eberlein (2006)]. There is also a *Dirichlet* η function $\eta(s)$, defined to be $(1 - 2^{1-s})\zeta(s)$, where ζ is the Riemann ζ function.

Theorem 5.7 (Ligozat [Ligozat (1975)], quoted in [McMurdy (2001)]). *Let N be a positive integer, and suppose that $f(z) = \prod_{0<\delta|N} \eta\left(q^\delta\right)^{r_\delta}$ is an η-quotient which satisfies the following properties:*

$$\sum_{0<\delta|N} \delta \cdot r_\delta \equiv 0 \mod 24 \tag{5.2}$$

and

$$\sum_{0<\delta|N} \frac{N}{\delta} \cdot r_\delta \equiv 0 \mod 24. \tag{5.3}$$

Then $f(z)$ satisfies

$$f\left(\frac{az+b}{cz+d}\right) = \chi(d)(cz+d)^k f(z) \tag{5.4}$$

for every $\left(\begin{smallmatrix} a & b \\ c & d \end{smallmatrix}\right) \in \Gamma_0(N)$, where $k := \frac{1}{2}\sum_{0<\delta|N} r_\delta$, and

$$\chi(d) := \left(\frac{(-1)^k \cdot s}{d}\right), \text{where } s := \prod_{0<\delta|N} \delta^{r_\delta}.$$

The function η is a useful building block for certain well-known modular forms and modular functions; it can be said to behave like a "modular form of weight $1/2$" (in some sense). In Section 5.3, we will use powers of η to find congruences that the Fourier coefficients of j satisfy. We can also write the Δ function in terms of η:

Theorem 5.8. *Let $z \in \mathcal{H}$ and $q := e^{\pi i z}$ as usual. Then the following equivalences hold:*

$$\Delta(q) = \sum_{n=1}^{\infty} \tau_n q^n = \eta(z)^{24} = q \prod_{n=1}^{\infty} (1-q^n)^{24}.$$

Proof. We will prove this by means of a lemma on the transformation properties of the η function. This will show that $\eta(z)^{24}$ transforms like a weight 12 modular form under the action of the matrix $\left(\begin{smallmatrix} 0 & -1 \\ 1 & 0 \end{smallmatrix}\right)$. We see that $\eta(z)^{24}$ does transform correctly under the action of $z \mapsto z+1$, so the lemma will show that $\eta(z)^{24}$ is weakly modular of weight 12. It is clearly holomorphic at ∞, as it has a Fourier expansion with no negative terms.

We also see that $\eta(z)$ converges to a nonzero value for any $z \in \mathcal{H}$, by considering the product expansion given. It therefore defines a holomorphic function on $\mathcal{H}$. This means that $\eta(q)^{24}$ is the Fourier expansion of a cusp form of level 1 and weight 12. This is a 1-dimensional space of modular forms, so we know that $\Delta = c \cdot \eta^{24}$ for some constant c. We note that both Δ and $\eta(q)^{24}$ have Fourier expansions which begin $q - 24q^2 + O(q^3)$, so therefore $c = 1$. $\square$

We now state and prove the lemma on the transformation properties of η.

Lemma 5.9. *Let* $\sqrt{}$ *be the branch of the square root function which has non-negative real part. Then we have*

$$\eta(-1/z) = \sqrt{z/i} \cdot \eta(z). \tag{5.5}$$

Proof. If we can show that the logarithmic derivatives of the left- and right-hand sides of (5.5) are equal, then we will have shown that this equation holds up to a multiplicative constant. However, if we substitute $z = i$ into this, then we see that the constant has to be 1, so if it holds it will be an identity.

If we take the logarithmic derivative of $\eta(z)$ as given in (5.1) with respect to z (remembering that $q = e^{2\pi i z}$), then we obtain

$$\frac{\eta'(z)}{\eta(z)} = \frac{2\pi i}{24}\left(1 - 24\sum_{n=1}^{\infty}\frac{nq^n}{1-q^n}\right).$$

We can rewrite the fraction inside the sum as a geometric series in q^n, and collect up the terms with a particular power of q to obtain

$$\frac{\eta'(z)}{\eta(z)} = \frac{2\pi i}{24}\left(1 - 24\sum_{n=1}^{\infty} n \sum_{m=1}^{\infty} q^{nm}\right) \tag{5.6}$$

$$= \frac{2\pi i}{24}\left(1 - 24\sum_{m=1}^{\infty}\sum_{n=1}^{\infty} nq^{mn}\right) \tag{5.7}$$

$$= \frac{2\pi i}{24}\left(1 - 24\sum_{m=1}^{\infty}\sigma_1(m)q^m\right) = \frac{2\pi i}{24}E_2(z). \tag{5.8}$$

We recall from Proposition 2.9 that the transformation formula for $E_2(z)$ is

$$E_2(-1/z)\cdot z^{-2} = \frac{12}{2\pi i z} + E_2(z);$$

We now substitute (5.8) into this to obtain

$$\frac{\eta'(-1/z)}{\eta(-1/z)}\cdot z^{-2} = \frac{1}{2z} + \frac{\eta'(z)}{\eta(z)}. \tag{5.9}$$

We see that this is the logarithmic derivative of the relation (5.5), which is what we wanted to prove, so we are done. ☐

In fact, one can derive a more general formula for the transformation of η under $\mathrm{SL}_2(\mathbf{Z})$: this is given by

$$\eta\left(\frac{az+b}{cz+d}\right) = \varepsilon\cdot(cz+d)^{1/2}\eta(z) \text{ for } \begin{pmatrix} a & b \\ c & d\end{pmatrix} \in \mathrm{SL}_2(\mathbf{Z}),$$

where ε depends on $\left(\begin{smallmatrix} a & b \\ c & d \end{smallmatrix}\right)$, and $\varepsilon^{24} = 1$. (The explicit definition of ε, which depends on a, b, c, d, is somewhat complicated, so we do not give it here, but it is given in Theorem 3.4 of [Apostol (1976)]). One can derive this formula for any given $\gamma \in \mathrm{SL}_2(\mathbf{Z})$ by writing γ in terms of the matrices S and T and using Lemma 5.9.

There are certain spaces of modular forms which are one-dimensional and spanned by η-products. These modular forms are eigenvectors for the Hecke operators and therefore their coefficients are multiplicative and are integral. The first example of this is the modular form Δ, which we have just shown to have an η-product expansion. The following result gives a list of some η-products which are eigenforms.

Theorem 5.10. *Let N and k be positive integers such that $k(N+1) = 24$. Then $\left(\eta(q) \cdot \eta(q^N)\right)^k \in S_k(\Gamma_0(N))$, unless $k = 1$ or $k = 3$, in which case we have $\left(\eta(q) \cdot \eta(q^N)\right)^k \in S_k(\Gamma_0(N), (\frac{\cdot}{N}))$.*

Proof. The pairs (k, N) which satisfy this criterion are $(12,1)$, $(8,2)$, $(6,3)$, $(4,5)$, $(3,7)$, $(2,11)$, and $(1,23)$. We have just shown that $\eta(q)^{24}$ is the unique normalized cusp form of weight 12 for $\mathrm{SL}_2(\mathbf{Z})$, so we have already proved the theorem in the $(12,1)$ case.

Let S be $S_k(\Gamma_0(N))$, if k is even, and $S_k(\Gamma_0(N), (\frac{\cdot}{N}))$, if k is odd. We define $g := \left(\eta(q) \cdot \eta(q^N)\right)^k$.

This proof will be in two parts. Firstly, we will show that if there is a nonzero modular form F in S, then it is a linear multiple of g. This does by itself not show that g is a modular form; it could happen that S is zero-dimensional, in which case g would not be a modular form. We then use the dimension formulae to show that, in fact, S is a 1-dimensional vector space.

The key idea in this proof is to consider the modular form

$$h := g^{N+1} = \left(\eta(q) \cdot \eta(q^N)\right)^{k(N+1)} = \Delta(q) \cdot \Delta(q^N) \tag{5.10}$$

$$= q^{N+1} \prod_{n=1}^{\infty} (1 - q^n)^{24} (1 - q^{Nn})^{24}. \tag{5.11}$$

Since $\Delta \in S_{12}(\Gamma_0(1))$, we see that $\Delta(q^N) \in S_{12}(\Gamma_0(N))$ (it is an oldform of level N, in the image of the map $q \mapsto q^N$) and therefore that $h \in S_{24}(\Gamma_0(N))$.

The Fourier expansion given in (5.11) shows us that h has a zero of multiplicity exactly $N+1$ at ∞; we note that it has no zeroes on $\mathcal{H}$, because Δ has no zeroes on $\mathcal{H}$.

We will now consider the Fourier expansion of h at the cusp 0; we notice that all of the possible N given in the list are *prime* (except 1, which we have already dealt with). From Exercise 6 of Chapter 2, we see if N is prime then the congruence subgroup $\Gamma_0(N)$ has exactly two cusps, which we may take to be 0 and ∞. Therefore, it will suffice to show that g has no poles at 0 and at ∞, and we have already seen that g does not have a pole at ∞, by considering its Fourier expansion at ∞.

To obtain its Fourier expansion at the cusp 0, we consider

$$\begin{aligned} h(z)|[\left(\begin{smallmatrix} 0 & -1 \\ 1 & 0 \end{smallmatrix}\right)]_{24} &= z^{-24}\Delta(-1/z)\cdot\Delta(-N/z) &(5.12)\\ &= z^{-24}\cdot z^{12}\cdot(z/N)^{12}\Delta(z)\cdot\Delta(z/N) &(5.13)\\ &= N^{-12}q_N^{N+1}\prod_{n=1}^{\infty}((1-q_N^n)(1-q_N^{Nn}))^{24}, &(5.14)\end{aligned}$$

where the transformation in (5.13) comes from the transformation formula for a modular form of weight 12, and $q_N := e^{(2\pi i z)/N}$. We see that h has a zero of exactly multiplicity $N+1$ in its Fourier expansion at the cusp 0.

Now we consider our modular form $F \in S$. Because this is a cusp form, it vanishes at every cusp, so it has a Fourier expansion at ∞ which is divisible by q and a Fourier expansion $F|[\left(\begin{smallmatrix} 0 & -1 \\ 1 & 0 \end{smallmatrix}\right)]_k$ which is divisible by q_N. Therefore F^{N+1} has a Fourier expansion at ∞ which is divisible by q^{N+1} and a Fourier expansion at 0 which is divisible by q_N^{N+1}.

We will now consider the quotient $(F/h)^{N+1}$; because both F^{N+1} and h^{N+1} are in $S_{24}(\Gamma_0(N))$, we see that their quotient is a modular function of weight 0 for $\Gamma_0(N)$. By the arguments above, we see that, as h has no zeroes on $\mathcal{H}$, the quotient is holomorphic on the upper half-plane. We now show that it is in fact a modular *form* of weight 0 for $\Gamma_0(N)$, as it does not have a pole at the cusps.

Because F^{N+1} has zeroes of multiplicity at least $N+1$ at both 0 and ∞, the zeroes of multiplicity exactly $N+1$ at 0 and ∞ that h^{N+1} has are cancelled out, so the quotient is indeed a modular form of weight 0. By Corollary 3.15, we see that this means that it must be a constant, which is what we needed to show.

Finally, we can check using the dimension formulae from Section 3.1 that $S_k(\Gamma_0(N))$ has dimension exactly 1. This shows that there does exist a nonzero modular form $F \in S$, so therefore F is a constant multiple of $(\eta(q)\eta(q^N))^k$. □

We note that the second part of the proof is necessary; we see that the argument given here shows that any nonzero modular form in $S_3(\Gamma_0(7))$

is a constant multiple of $(\eta(q)\eta(q^7))^3$, but we have already seen that this space must have dimension 0 (because the weight is odd and $\left(\begin{smallmatrix}1&1\\0&1\end{smallmatrix}\right) \in \Gamma_0(7)$). It can be shown that the spaces of modular forms with character mentioned in the theorem are one-dimensional, and then the same proof will show that every modular form in these spaces is a linear multiple of $(\eta(q)\eta(q^N))^k$.

In [Shimura (1994)], the following result is stated as Exercise 2.29:

Theorem 5.11. *Let* $N \in \{2, 3, 4, 6, 12\}$, *and let* $k = 12/N$. *Then*

$$S_k(\Gamma(N)) = \mathbf{C} \cdot \Delta(z)^{1/N}.$$

Remark 5.12. These modular forms are interesting because they are truly modular forms for $\Gamma(N)$ and not for a subgroup of the form $\Gamma_1(M)$ for any integer M. This follows by inspection because they do not have a Fourier expansion in terms of q, but instead in terms of q_N.

Proof. [Proof of Theorem 5.11] We leave the proof as Exercise 3. There are two things that we need to show; first, that the spaces of modular forms are all 1-dimensional, and second that the η-products given are actually modular forms for the given $\Gamma(N)$.

The first of these can be derived from the formulae for the dimension of spaces of modular forms given in Chapter 3, which also requires us to work out the index of $\Gamma(N)$ in $\mathrm{SL}_2(\mathbf{Z})$, as we did in Section 2.4.

One way to prove the second assertion is to find a set of matrices which generate $\Gamma(N)$ for each of the N in the theorem, and then to show that $\Delta(z)^{1/N}$ transforms correctly under each of them. We can use the discussion of $\mathbf{P}^1(\mathbf{Z}/N\mathbf{Z})$ from Section 2.4 to find suitable generators for $\Gamma(N)$, and then show that our η-product transforms correctly under these. □

5.3 The arithmetic of the j-invariant

In Section 5.1, we introduced the j-invariant as a modular function of weight 0 for $\mathrm{SL}_2(\mathbf{Z})$. We will now consider its arithmetic theory, using the properties of the η function that we have derived.

Firstly, we note explicitly that, because j is a modular function of weight 0 for $\mathrm{SL}_2(\mathbf{Z})$, it satisfies the transformation properties

$$j(z+1) = j(z) \text{ and } j\left(-z^{-1}\right) = j(z), \text{ for } z \in \mathcal{H}.$$

(We note that the term in $(cz + d)$ vanishes because the weight is 0).

We now define the Fourier coefficients $c(n)$ of the j-invariant by

$$j(q) = \frac{1}{q} + \sum_{n=0}^{\infty} c(n)q^n \tag{5.15}$$

$$= \frac{1}{q} + 744 + 196883q + 21493760q^2 + O(q^3). \tag{5.16}$$

If we compute a few of these $c(n)$, by using the definition of the j-invariant as E_4^3/Δ, we see that they are integers. This is in fact true in general; we have

Proposition 5.13. *The $c(n)$ we defined in* (5.15) *are all integers.*

Proof. We follow the proof given in Section 1.15 of [Apostol (1976)], which relies heavily on the fact that j is defined as a quotient of two power series whose Fourier coefficients are both integral. Let P and Q denote power series with integral coefficients. We have that

$$j(q) = \frac{E_4(q)^3}{\Delta(q)} = \frac{1+P}{q(1+Q)}$$
$$= \frac{1}{q} \cdot (1+P) \cdot (1+Q)^{-1},$$

so the Fourier expansion of the j-invariant has integral coefficients. □

These Fourier coefficients have arithmetic significance; for instance, if α is a positive integer then Lehner proved in [Lehner (1949a)] and [Lehner (1949b)] that the following congruences hold:

$$c(2^\alpha n) \equiv 0 \mod 2^{3\alpha+8}$$
$$c(3^\alpha n) \equiv 0 \mod 3^{2\alpha+3}$$
$$c(5^\alpha n) \equiv 0 \mod 5^{\alpha+1}$$
$$c(7^\alpha n) \equiv 0 \mod 7^{\alpha}$$
$$c(11^\alpha n) \equiv 0 \mod 11^{\alpha}.$$

Similarly, Lehmer proved that

$$(n+1)c(n) \equiv 0 \mod 24, \text{ for } n \geq 1.$$

Chapter 4 of Apostol's book on modular functions [Apostol (1976)] gives a good introduction to this topic.

There are congruences of a different form for the prime 13; it is a theorem of Newman [Newman (1958)] that

$$c(13np) + c(13n) \cdot c(13p) + p^{-1}c\left(\frac{13n}{p}\right) \equiv 0 \mod 13.$$

We can tell that the congruences for 13 are going to be of a different form than those for the smaller primes, because $c(13) \not\equiv 0 \mod 13$ (and indeed $c(13^2) \not\equiv 0 \mod 13$ as well).

Here, we quote and leave as an exercise Theorem 4.11 from [Apostol (1976)], which will prove the $\alpha = 1$ case of the congruences above.

Theorem 5.14. *Let $p \in \{2, 3, 5, 7, 13\}$, and define*

$$\Phi(q) := \left(\frac{\eta(q^p)}{\eta(q)} \right)^s, \quad \textit{where } s = \frac{24}{p-1}.$$

There exist integers $a_1, \ldots, a_{p^2}$ such that

$$j_p(q) := \sum_{n=1}^{\infty} c(pn)q^n = p^{s/2-1} \cdot \left(a_1 \Phi(q) + \cdots + a_{p^2} \Phi(q)^{p^2} \right). \tag{5.17}$$

Proof. See Exercise 10; we multiply both sides of (5.17) by carefully chosen modular forms (with zeroes that cancel the poles of the modular functions involved) to turn this into a question about an equality of modular *forms*, which can be solved by a finite computation and an application of the Sturm bound. □

Remark 5.15. We notice that the list of primes p in Theorem 5.14 are precisely those for which $S_2(\Gamma_0(p)) = 0$; this is related to the subject of *modular curves.* These are Riemann surfaces which are constructed as quotients of the upper half-plane $\mathcal{H}$ by the action of congruence subgroups, and in the case when this space of cusp forms has dimension 0, the corresponding modular curve has genus 0. This fascinating subject is unfortunately beyond the scope of this book. A good introduction to this subject is [Diamond and Im (1995)].

We note that there is an equality for $p = 13$ given by Theorem 5.14, but it can be seen that the exponent $s/2 - 1$ is 0 when $p = 13$, so there is no nontrivial congruence given by this formula for $p = 13$.

If we explicitly compute the $c(n)$, then we see that they are growing quite rapidly. In fact, we have the following equation (proved by Petersson in [Petersson (1932)]):

$$c(n) \approx \frac{e^{4\pi\sqrt{n}}}{\sqrt{2}n^{3/4}}. \tag{5.18}$$

We will exhibit formulae for the growth of the Fourier coefficients of modular forms later in this chapter; we will see that the coefficients of $\Delta(q)$, for instance, are of a different order of magnitude.

5.3.1 *The j-invariant and the Monster group*

It is a standard exercise in undergraduate group theory courses to show that the alternating group on five symbols, A_5, is the smallest non-abelian simple group; in other words, it is the smallest non-abelian group with no nontrivial normal subgroups (see [David (1987)] for an elementary two-page proof of this). During the late 20th and early 21st centuries, a great deal of work went into classifying all of the finite simple groups; this theorem has been called "the enormous theorem", due to its extreme length (about 15000 pages). We note that there were some questions raised about the proof, due to it being spread over many journals and written by many people, and in recent years a more coherent and organized version of the proof has been created; see the series of volumes beginning with [Gorenstein *et al.* (1994)].

This classification of finite simple groups tells us that there are 18 countably infinite families of simple groups (A_5 falls into one of these), and 26 other simple groups (the "sporadic groups") which fall outside these families. The largest of these 26 groups is known as the *Monster Group.* It has order $\backsim 8 \cdot 10^{53}$, and was first constructed as the automorphism group of a certain 196884-dimensional commutative nonassociative algebra, which has a 196883-dimensional faithful representation.

We note that $c(1) = 196883$; this is not a numerological coincidence, but follows from a branch of mathematics called *moonshine theory* which arises from string theory, and relates the Monster group and modular functions. Borcherds's proof of this connection and others in [Borcherds (1992)] earned him the Fields Medal in 1998. We refer to the survey article [Gannon (2006)] for more details. It had also been noticed by Ogg that the primes dividing the order of the Monster group are exactly those for which $S_2(\Gamma_0(p)+)$ has dimension 0, where $\Gamma_0(p)+$ is the group generated by $\Gamma_0(p)$ and $\left(\begin{smallmatrix} 0 & -1 \\ p & 0 \end{smallmatrix}\right)$ (we note that because this group is bigger than $\Gamma_0(p)$ the space of modular forms for it is smaller).

Borcherds proved the connection by proving the Conway-Norton Conjecture (see [Conway and Norton (1979)]), which relates the coefficients of certain weight 0 modular functions (such as the j-invariant) with linear combinations of the dimensions of certain irreducible representations of the Monster group; the simplest example is that for $c(1)$ which we gave above, which was an observation of McKay (see Observation (F) of [Conway and Norton (1979)]). This is a good example of how computations can guide the development of theory.

5.3.2 *"Ramanujan's Constant"*

We now pass from the highly sophisticated and important results of Borcherds on modular moonshine to consider an area whose name is derived from an April Fools' Day joke.

An interesting arithmetical coincidence arises from the following fact. The j-invariant is a modular function, so it is in particular a function from $\mathcal{H}$ to $\mathbf{C}$. It is a theorem (Theorem 6.1 and Example 6.2.1 of [Silverman (1994)]) that if the quadratic imaginary field $K = \mathbf{Q}(\sqrt{d})$ has class number 1, and the ring of integers $\mathcal{O}_K$ is generated over $\mathbf{Z}$ by $\{1, \alpha\}$, then $j(\alpha) \in \mathbf{Z}$. It is well-known that $\mathbf{Q}(\sqrt{-163})$ has class number 1, and that its ring of integers is generated by $\left\{1, \frac{1+\sqrt{-163}}{2}\right\}$, so therefore

$$j\left(\frac{1+\sqrt{-163}}{2}\right) \in \mathbf{Z}.$$

We will now see that the Fourier expansion of the j-invariant converges very quickly in this case. If we let $z = (1+\sqrt{-163})/2$, then we find that

$$q = e^{2\pi i z} = -e^{-\pi\sqrt{163}} \approx -3.809 \times 10^{-18}.$$

This is a very small number, so the dominant terms in the Fourier expansion of j are going to be $1/q + 744$. We see that $1/q$ is going to be very close to an integer. In fact, if we calculate it we find that

$$e^{\pi\sqrt{163}} = 262537412640768743.99999999999925\ldots, \tag{5.19}$$

so to twelve decimal places, it appears to be an integer. It is in fact a transcendental number, by the Gelfond-Schneider theorem (see [Baker (1990)], Chapter 2, for more details on this). Similar results hold for the nine integers d such that $\mathbf{Q}(\sqrt{-d})$ has class number 1.

This number $e^{\pi\sqrt{163}}$ is sometimes called "Ramanujan's Constant"; this is a misnomer, as it was actually discovered by Hermite in the 1850s [Hermite (1859)], and was in fact not one of the formulae given by Ramanujan in his paper on the subject [Ramanujan (1913–1914)]. The name was in fact coined in the 1970s, following an April Fool's joke played by Martin Gardner[2]. One formula that Ramanujan did give in his paper was

$$e^{\pi/4\sqrt{102}} \approx 800\sqrt{3} + 196\sqrt{51},$$

which is accurate to nine decimal places. Similar formulae can be derived using knowledge of class numbers of quadratic fields; see Example 6.2.2 of [Silverman (1994)].

[2]He claimed that it was an integer in his column.

In Exercise 11 we will find some similar formulae for what are sometimes known as "almost integers". This is not a precisely-defined term; it refers to real numbers like $e^{\pi\sqrt{163}}$ which are closer than "one would expect" to being integers.

There are other formulae involving what are known as the "singular moduli" of the j-invariant (that is, the values of the j-invariant on particular arguments which generate lattices in $\mathbf{C}$). We will not discuss these in detail in this book, but refer to [Cox (1989)], Section 12, for the proof of the following result of Weber:

$$j(\sqrt{-14}) = 2^3\left(323 + 228\sqrt{2} + (231 + 161\sqrt{2})\sqrt{2\sqrt{2}-1}\right)^3.$$

(We note that we can check this equality to 4 decimal places of accuracy just by using the first two terms of the Fourier expansion of the j-invariant).

5.4 Applications of the modular function $\lambda(z)$

We have now seen the j-invariant, one of the most commonly seen modular functions, and considered its arithmetic and its applications. In this section, we will define another modular function, and see how it can be used to compute digits of π. A reference for this section is Chapter 6 of [Armitage and Eberlein (2006)].

We define $\lambda(z)$ by

$$\lambda(z) := \left(\frac{\sum_{n=-\infty}^{\infty} q^{((n+1)/2)^2}}{\sum_{n=-\infty}^{\infty} q^{n^2}}\right)^8;$$

this can be shown to be a modular function of weight 0 for $\Gamma(2)$ (see Exercise 5, as it has the following η-quotient expansion:

$$\lambda(q) = \left(\frac{\eta(\sqrt{q})\eta(q^2)^2}{\eta(q)^3}\right)^8. \tag{5.20}$$

We note again the ubiquity and usefulness of the Dedekind η function as a tool to create new modular forms and functions.

This function is also sometimes known in the literature as k^2, k being the name of another modular function. Given the universal use of the letter k in modern works for the weight of a modular form or function, this notation is probably best deprecated[3].

[3] The late Serge Lang, author of many mathematics book, including [Lang (1976)], was

We note that in [Littlewood (1986)] λ (under the name k^2) is known as "the *modular* function which arises from the theory of elliptic functions", as opposed to "the elliptic modular function", which can be either λ or j in the literature.

The λ function is related to the j-invariant by the following formula:

$$j(q) = 256 \cdot \frac{(1-\lambda(q)+\lambda(q)^2)^3}{\lambda(q)^2 \cdot (1-\lambda(q))^2}. \tag{5.21}$$

We can check this by turning this into a relation between modular *forms* by multiplying both sides by a suitable modular form (to cancel the poles), and then using the Sturm bound to verify the identity.

Either by explicitly computing $\lambda(q)$ or by dimensional analysis of (5.21), we can see that the Fourier expansion of λ at ∞ is in terms of q_2 rather than q, which indicates that it is genuinely a modular function for $\Gamma(2)$, rather than a larger congruence subgroup.

5.4.1 *Computing digits of π using $\lambda(z)$*

The calculation of decimal digits of π has been a goal of mathematicians since classical times. Archimedes obtained the following bound in the 3rd century BC by considering polygons containing and contained by a circle:

$$3\frac{10}{71} < \pi < 3\frac{1}{7};$$

this line of enquiry was followed until the 17th century AD, when Ludolph van Ceulen computed π to 34 digits by considering polygons with approximately 2^{60} sides. The number π was known as *the Ludolphine number* in Germany for a long time after his death in his honour.

Gregory's formula for π, which is

$$\frac{\pi}{4} = \arctan(1) = 1 - \frac{1}{3} + \frac{1}{5} - \frac{1}{7} + \cdots$$

is very beautiful and simple, but unfortunately does not converge very quickly; in fact, because the error after the nth term is greater than $1/(2n)$, taking 300 terms of the series does not even give the first two decimal

a scourge of those who used what he considered bad notation, often using the phrase "this notation sucks!". Once Barry Mazur tried deliberately to get Lang to say this, by introducing a variable called Ξ (written as three parallel lines), then dividing this by its complex conjugate, thus giving eight parallel horizontal lines; this notation is worse than any that is given in this book [Jorgenson and Krantz (2006)].

digits correctly. A much more efficient way of computing π is to use several trigonometric series expansions, such as the following formula due to Machin:

$$\frac{\pi}{4} = 4 \cdot \arctan\left(\frac{1}{5}\right) - \arctan\left(\frac{1}{239}\right); \tag{5.22}$$

Machin's formula was used by Shanks in the 1850s to calculate 707 digits of π by hand; although these figures were accepted for nearly a century, it was discovered in the 1940s that there was an error in the 527th place, and all of the subsequent digits were incorrect.

There is a third type of algorithm, which use the properties of the λ function. We refer to the survey article [Borwein *et al.* (1989)] for an example of this type.

Algorithm 5.16 (Algorithm 1 of [Borwein *et al.* (1989)]). *Let $\alpha_0 := 6 - 4\sqrt{2}$ and $y_0 := \sqrt{2} - 1$. For $n \in \mathbf{N}$, define*

$$y_{n+1} := \frac{1 - (1 - y_n^4)^{1/4}}{1 + (1 - y_n^4)^{1/4}}, \tag{5.23}$$

$$\alpha_{n+1} := (1 + y_{n+1})^4 \alpha_n - 2^{2n+3} y_{n+1}(1 + y_{n+1} + y_{n+1}^2). \tag{5.24}$$

Then we have

$$0 < \alpha_n - \frac{1}{\pi} < 16 \cdot 4^n \cdot e^{-2 \cdot 4^n \pi},$$

and α_n converges to $1/\pi$ quartically; in other words, every iteration quadruples the number of correct decimal digits.

Although this algorithm is *not* due to Ramanujan, it should be noted that he did consider using modular equations to approximate π in his paper [Ramanujan (1913–1914)].

The complete proof that this algorithm works requires a discussion of elliptic integrals which is beyond the scope of this book, but we can sketch some of the main ideas. We refer to [Borwein *et al.* (1989)] for an overview of this, and to the book [Borwein and Borwein (1987)] for a more detailed account.

We can write the elliptic integrals in terms of modular functions (this is why λ is sometimes called the *elliptic* modular function in the literature), and we can define a function α (which will supply us with the α_n in the algorithm) by

$$\alpha(r) := \frac{\frac{1}{\pi} - 4\sqrt{r} \cdot \frac{\sum_{n=-\infty}^{\infty} n^2 (-s)^{n^2}}{\sum_{n=-\infty}^{\infty} (-s)^{n^2}}}{\left(\sum_{n=-\infty}^{\infty} s^{n^2}\right)^4},$$

where $s = e^{-\pi\sqrt{r}}$.

We see that this series is closely related to that of λ; it is obtained by considering functions of λ and its derivative. We can see that as $r \to \infty$ $s \to 0$ and $\alpha(r)$ tends to $1/\pi$. The following approximation can be shown to hold:

$$\alpha(r) - \frac{1}{\pi} \approx 8\left(\sqrt{r} - \frac{1}{\pi}\right) \cdot e^{-\pi\sqrt{r}};$$

one then obtains a good approximation to $1/\pi$, and hence to π, by finding an iterative way to approximate $\alpha(r)$. This can be provided by

$$\alpha(25r) = t^2\alpha(r) - \sqrt{r} \cdot \left(\frac{t^2-5}{2} + \sqrt{t(t^2-2t+5)}\right),$$

where $t := \sum_{n=-\infty}^{\infty}(s^5)^{n^2} / \sum_{n=-\infty}^{\infty} s^{n^2}$. We see again, by the shape of this relationship, that this is related to the theory of modular functions. This provides a good example of a genuine application of modular functions to the wider world of mathematics; although this book is mainly concerned with modular forms and number theory, it should not be forgotten that modular forms are also complex-analytic objects.

5.4.2 *Proving Picard's Theorem*

In Littlewood's *A Mathematician's Miscellany* [Littlewood (1986)][4], the question of how short a PhD thesis in mathematics could be is discussed (the question had arisen in conversation). Littlewood argued that there are results which can be stated in one line and proved in a second, which were of a quality to be worth a PhD. He gave as an example Picard's (Little) Theorem:

Theorem 5.17 ("Picard's Little Theorem"). *Let f be an entire function which misses two distinct values. Then f is constant.*

Proof. [We will need to know that λ is a 1-1 function on a fundamental domain F for $\Gamma(2)$.]

Let f be an entire function that is never 0 or 1. Then $\exp(\lambda^{-1}(f(z)))$ is a *bounded* entire function, and is therefore constant. □

Here the line in brackets is not counted in the "two lines" requirement[5].

[4] To quote Littlewood, "A Miscellany is a collection without a natural ordering relation".

[5] This is very reminiscent of the famous anonymous quote that "[t]his is a one line proof ... if we start sufficiently far to the left."

We note that there does exist a function which is entire and misses exactly one point; this is $\exp(z)$, which is never 0; the condition given in the theorem is necessary.

5.5 Identities of series and products

A classical application of modular forms is to find equalities of arithmetical functions, by finding modular forms whose Fourier coefficients are those arithmetical functions, and then finding identities of modular forms which follow from dimensional considerations. For instance, the following well-known identities of multiplicative functions hold:

$$\sigma_7(n) = \sigma_3(n) + 120 \sum_{i=1}^{n-1} \sigma_3(i) \cdot \sigma_3(n-i), \tag{5.25}$$

$$11\sigma_9(n) = 21\sigma_5(n) - 10\sigma_3(n) + 5040 \sum_{i=1}^{n-1} \sigma_3(i) \cdot \sigma_5(n-i), \tag{5.26}$$

$$\sigma_{13}(n) = 21\sigma_5(n) - 20\sigma_7(n) + 10080 \sum_{i=1}^{n-1} \sigma_5(i) \cdot \sigma_7(n-i). \tag{5.27}$$

Proving these identities is left as Exercise 7. We note that these can be proved without using modular forms, but once we have set up enough of the theory of modular forms (especially their dimensions) results like this follow almost immediately.

We can also use the results of Section 5.2 to find identities between Fourier series defined as sums and Fourier series defined as products. Using the definition of Δ in terms of E_4 and E_6, we have

$$\Delta(q) = \sum_{n=1}^{\infty} \tau(n) q^n \tag{5.28}$$

$$= q \prod_{n=1}^{\infty} (1 - q^n)^{24} \tag{5.29}$$

$$= \frac{\left((1 + 240 \sum_n \sigma_3(n) q^n)^3 - (1 - 504 \sum_n \sigma_5(n) q^n)^2\right)}{1728}; \tag{5.30}$$

this follows directly from the definitions and the dimensional consideration. Similar identities hold for the coefficients of the cusp forms for $\mathrm{SL}_2(\mathbf{Z})$ of weights 16, 18, 20, 22 and 26, where the same dimensional constraints hold, and also for congruence subgroups where the dimensional considerations imply such results; we refer to [Emmons (2005)] for similar results classifying all eigenforms for which simple results like those of this section hold.

We can also prove congruence relations between certain coefficients of modular forms; for instance, we have that

$$\tau(n) \equiv \sigma_{11}(n) \mod 691,$$

which relates the Fourier coefficients of Δ and E_{12}; this is Exercise 12. We can show this and similar results by finding linear dependencies between modular forms with integral Fourier expansions and then reducing these modulo a suitable prime p.

The study of congruences between modular forms is an integral part of the modern theory of modular forms; one can create "families" of modular forms that have Fourier expansions which are congruent to one another modulo powers of primes. The first example of this are sets of Eisenstein series $\{E_{k+(p-1)p^n}\}$, which we can show have Fourier coefficients which are congruent to $E_k(q)$ modulo p^n; see Exercise 14 (and Proposition 6.7 for the necessary result on the constant term).

5.6 The Ramanujan-Petersson Conjecture

One of the many things of arithmetic interest in the theory of modular forms is the magnitude of the absolute value of the Fourier coefficients of a modular form. We see that the Fourier coefficients of the Eisenstein series E_k satisfy $O(n^{k-1})$, from the definition of $\sigma_{k-1}(n)$. It is natural to ask what the corresponding formulae should be for normalized cuspidal modular eigenforms, and for other non-cuspidal modular forms.

There is a result of Hecke on the size of the coefficients of a cusp form which we will state and prove (following the exposition in [Serre (1973a)], Theorem VII.5). The proof of the sharper result that is the Ramanujan-Petersson conjecture is beyond the scope of this book; we will merely state it.

Theorem 5.18 (Hecke). *Let f be a nonzero cuspidal modular form for $\mathrm{SL}_2(\mathbf{Z})$ of weight k with Fourier expansion $f(q) = \sum_{n=1}^{\infty} a_n q^n$. Then $a_n = O(n^{k/2})$.*

Proof. We note that, as f is a cusp form, it vanishes at ∞, so we can factor q out of the Fourier expansion of f. This means that we have

$$|f(z)| = O(q) = O(e^{-2\pi y}) \text{ with } y = \Im(z), \text{ as } q \to 0. \tag{5.31}$$

Let $g(z) = |f(z)| \cdot y^{k/2}$. It is an elementary exercise (Exercise 4) to show that $\Im(\gamma(z)) = \Im(z)/(|cz+d|^2)$ and therefore that $g(z)$ is invariant under the action of $\left(\begin{smallmatrix} a & b \\ c & d \end{smallmatrix}\right) \in \mathrm{SL}_2(\mathbf{Z})$.

We also see that g is continuous on the standard fundamental domain D and that it tends to 0 as $y \to \infty$. This means that g is bounded; that is, there exists a constant M such that

$$|f(z)| \leq M \cdot y^{-k/2} \text{ for all } z \in \mathcal{H}. \tag{5.32}$$

Let $z = x + iy$. We fix y and vary x between 0 and 1; this means that $q = e^{2\pi i(x+iy)}$ runs anticlockwise along a circle C_y with centre at 0. This means that, by the residue formula,

$$a_n = \frac{1}{2\pi i}\int_{C_y} f(z) \cdot q^{-n-1} dq = \int_0^1 f(x+iy) q^{-n} dx.$$

We now use the estimate of (5.32) to obtain the following estimate for $|a_n|$:

$$|a_n| \leq M \cdot y^{-k/2} \cdot e^{2\pi n y}.$$

This is valid for every $y > 0$. We will therefore choose $y = 1/n$; this will make our inequality for $|a_n|$ independent of y, and therefore we have

$$|a_n| \leq e^{2\pi} \cdot M \cdot n^{k/2};$$

we see that $a_n = O(n^{k/2})$ as required. □

We note that the implied constant in the $O(n^{k/2})$ will depend on the first few coefficients of f; we will see that the sharp bounds given by the Ramanujan-Petersson conjecture assume that we are dealing with a normalized eigenform.

We will now show that the Fourier coefficients of a non-cuspidal modular form for $\mathrm{SL}_2(\mathbf{Z})$ are of size $O(n^{k-1})$.

Corollary 5.19. *Let f be a non-cuspidal modular form for $\mathrm{SL}_2(\mathbf{Z})$ of weight k with Fourier expansion $f(q) = \sum_{n=0}^{\infty} a_n q^n$. Then $a_n = O(n^{k-1})$.*

Proof. We can write $f = \lambda E_k + h$ with $\lambda \neq 0$ and h a cusp form (Exercise 5). We now use the bounds we have on the coefficients of cuspidal modular forms and on the coefficients of E_k; we see that $n^{k/2}$ is much smaller than n^{k-1}, so the coefficients of E_k dominate, and therefore $a_n = O(n^{k-1})$ as required. □

Remark 5.20. We note that the coefficients of cuspidal and non-cuspidal modular forms differ noticeably in magnitude; it is often possible to tell them apart "by eye". The reader is encouraged to do experiments with their favourite computer algebra system to verify this.

By performing explicit computations, one can see that a somewhat smaller bound than that of Hecke appears to be true; this was conjectured by Ramanujan and Petersson, and proved by Deligne:

Theorem 5.21 (Ramanujan-Petersson Conjecture). *Let N be a positive integer and let f be a normalized cuspidal eigenform of weight k for $\Gamma_1(N)$ with Fourier expansion $f(q) = \sum_{n=1}^{\infty} a_n q^n$. Then we have that*

$$|a_n| \leq n^{(k-1)/2} \cdot \sigma_0(n).$$

Proof. This was proved by Deligne as a (deep) consequence of his proof of the Weil conjectures; we refer the reader to his seminal paper, [Deligne (1974)], which is described in its MathSciNet review as "without question the most important paper in algebraic geometry to have appeared in the last ten years". □

In particular, if p is prime then we see that the conjecture tells us that

$$|a_p| \leq 2p^{(k-1)/2}.$$

This result was motivated by study of the Fourier expansion of the Δ function; the conjecture in that case is that $\tau_p \leq 2p^{11/2}$; this specific case was conjectured by Ramanujan, and is therefore known as *Ramanujan's conjecture.*

We notice also that if $k = 1$ then the Ramanujan-Petersson conjecture tells us that the a_p are *bounded.* If we consider the (unique) normalized eigenform $f = \eta(q)\eta(q^{23}) \in S_1(\Gamma_1(23))$ then we see that

$$f(q) = q - q^2 - q^3 + q^6 + q^8 - q^{13} - q^{16} + q^{23} - q^{24} + O(q^{25})$$

and from the Ramanujan-Petersson conjecture it follows that $|a_p| \leq 2$; it can be shown that the values of a_p in fact lie in $\{-1, 0, 1, 2\}$. The first p such that $a_p = 2$ is 59.

One unsolved question about the Fourier coefficients of the Δ function was raised by Lehmer. It is simple to state:

Conjecture 5.22 (Lehmer). *Let n be a positive integer. Then $\tau_n \neq 0$.*

Lehmer showed that τ_n was nonzero for $n \leq 10^{15}$ [Lehmer (1947)], and this has been extended to $n \leq 2 \cdot 10^{19}$ in [Bosman (2007)], but the general case is still open. A similar question can be asked for the cusp forms of level 1 and weights 16, 18, 20, 22 and 26, where the space of cusp forms is 1-dimensional.

We will see in Section 5.10 examples of modular forms which are very far from satisfying the equivalent of Lehmer's conjecture; they are called *lacunary* modular forms, and have very large numbers of zero Fourier coefficients.

5.7 Elliptic curves and modular forms

The theory of elliptic curves has recently become very relevant to the theory of modular forms, with the proof of Fermat's Last Theorem, which we will discuss below.

There are several good books which discuss the theory of elliptic curves[6]; the two books of Silverman on the arithmetic of elliptic curves are standard references [Silverman (1992, 1994)], and some of the standard modular forms references are also good references for this subject (such as [Koblitz (1993)] and [Hellegouarch (2002)]). [Cassels (1991)], [Husemöller (2004)], [Knapp (1992)] and [Lang (1978)] are other good references for the theory of elliptic curves.

Definition 5.23. Let K be a field of characteristic not 2 or 3. We say that the set of projective solutions to

$$E : y^2 z = x^3 + a \cdot xz^2 + b \cdot z^3, \text{ with } a, b \in K \text{ and } 4a^3 + 27b^2 \neq 0 \tag{5.33}$$

is an *elliptic curve (in Weierstrass normal form).* We call the integer $\Delta(E) := 4a^3 + 27b^2$ the *discriminant* of E.

We recall from the discussion of projective space in Chapter 2 that there is a projective point which cannot be represented on a graph such as Figure 5.1, known as the point at infinity. In this formulation, it is the point $(0 : 1 : 0)$; this is sometimes viewed as being "infinitely far up the y-axis". [Buhler (2001)], Example 1.15, gives a good explanation of why the equation of E has this form. In sophisticated language, we say that an elliptic curve is a genus one curve with a defined point (the last part of this is non-trivial; there exist genus one curves without a rational point, such as the Selmer curve $3x^3 + 4y^3 + 5z^3 = 0$ over $\mathbf{Q}$, and these are not elliptic curves).

We note that there are slightly more complicated equations[7] which are valid for fields of characteristic 2 or characteristic 3.

[6]The Foreword to [Lang (1978)] begins "[i]t is possible to write endlessly on elliptic curves. (This is not a threat.)".

[7]The equation $y^2 + a_1 xy + a_3 y = x^3 + a_2 x^2 + a_4 x + a_6$ will work in any characteristic.

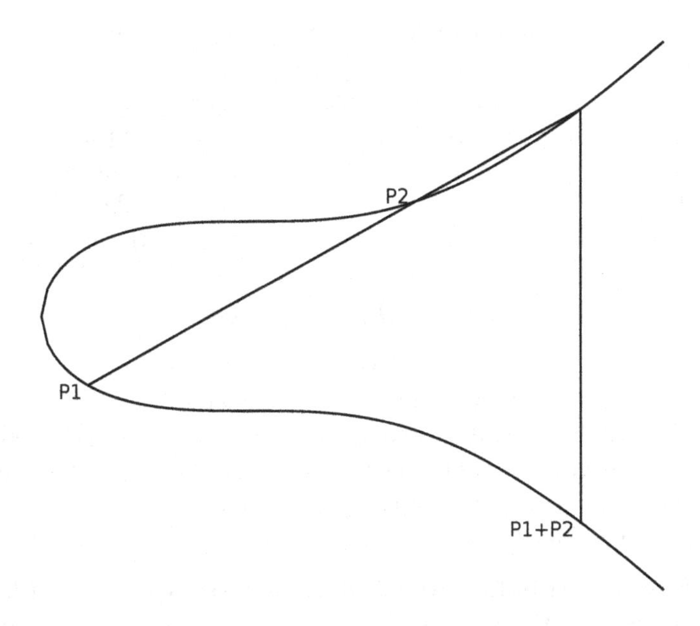

Fig. 5.1 Elliptic curve addition

We can give an elliptic curve a group law (which is abelian) by the procedure indicated in Figure 5.1 (this is known as the "chord-and-tangent method"; a version of this was probably known to Diophantus, and both Fermat and Newton developed this further), and there are some standard structure theorems for elliptic curves over number fields, which we will state here. The Mordell-Weil theorem tells us that the group $E(K)$ is finitely generated, so we have that

$$E(K) \cong \mathbf{Z}^r \oplus E_{\text{tors}},$$

where r is a non-negative integer and E_{tors} is called the *torsion subgroup*; this is the group of all elements with finite order. By work of Mazur [Mazur (1978)], it can be shown that the group $E_{\text{tors}}(\mathbf{Q})$ is finite (and in fact it is one of 15 groups, the largest of which is isomorphic to $\mathbf{Z}/2\mathbf{Z} \oplus \mathbf{Z}/8\mathbf{Z}$), and it is conjectured that for any given number field K we have

$$|E(K)_{\text{tors}}| \leq B(K),$$

where $B(K)$ is an integer depending only on K (and in particular is independent of the choice of E).

The example illustrated in Figure 5.1 is the addition of the points $(-2,-3)$ and $(2,5)$ to obtain $(4,-9)$ on the curve $E : y^2 = x^3 + 17$. This is given in more detail as Example 2.4 of [Silverman (1992)].

There is a reduction map; given an elliptic curve over $\mathbf{Q}$, we can choose its coefficients a and b to be integral, by multiplying through by a sufficiently large integer to clear the denominators of the coefficients and then applying a birational transformation to return the elliptic curve to Weierstrass normal form. The reduction map is given by

$$\begin{aligned} E &\to E(\mathbf{F}_p) := E \mod p \\ (x,y) &\mapsto (\overline{x},\overline{y}), \end{aligned}$$

where $\overline{x}$ and $\overline{y}$ are the reductions of x and y modulo p respectively. We note that $E(\mathbf{F}_p)$ will be an elliptic curve if and only if $4\overline{a}^3 + 27\overline{b}^2 \not\equiv 0$ modulo p.

One of the greatest results in modern number theory, which relates elliptic curves and modular forms, can be summed up as *every elliptic curve over* $\mathbf{Q}$ *is modular.* We now make more precise what this means.

Definition 5.24 (Modularity of elliptic curves over Q). *Let E be an elliptic curve over $\mathbf{Q}$ and let p be a prime such that $p \nmid \Delta(E)$. We define $a_p(E)$ to be*

$$a_p(E) := p + 1 - |E(\mathbf{F}_p)|.$$

We say that E is modular *if the a_p are the pth Fourier coefficients of a nonzero normalized cuspidal modular eigenform of weight 2.*

This was first raised by Taniyama and Shimura in the 1950s; Weil also considered this problem. It has been called the Shimura-Taniyama-Weil Conjecture, and many variations on this (Taniyama-Weil, Shimura-Taniyama, ...). A natural neutral name is the Modularity Conjecture.

Theorem 5.25 (The Modularity Theorem). *Let $E/\mathbf{Q}$ be an elliptic curve. Then E is modular.*

Remark 5.26. The work of Wiles and Taylor in [Wiles (1995)] and [Taylor and Wiles (1995)] proved the Modularity Conjecture in the "semistable case". Throughout the 1990s, work continued to prove the conjecture under less and less restrictive conditions[8]; first it was shown that the elliptic curve

[8] It has been noted that papers in the field from this period can be dated according to the version of the modularity theorem that they state, and the conditions they need to impose on the conductor.

needed only to be semistable away from 3 and 5 [Diamond (1996)], then that the conjecture was true if the conductor (see below) was not divisible by 27 [Conrad *et al.* (1999)]. Finally, [Breuil *et al.* (2001)] tackled the last case.

Remark 5.27. There is an algorithm to tell you what the level of the modular form associated to the elliptic curve E by Theorem 5.25 is; there is an integral invariant of an elliptic curve $E/\mathbf{Q}$ called the *conductor* which tells you when the curve $E/\mathbf{F}_p$ fails to be an elliptic curve; a method to compute this due to Tate is given in detail in [Silverman (1994)], Section 4.9. The Modularity Theorem says that the level of the modular form f_E attached to E is equal to the conductor of E. We say that an elliptic curve is *semistable at* p if its conductor is not divisible by p^2, and that it is *semistable* if the conductor is squarefree.

5.7.1 *Fermat's Last Theorem*

The work of Wiles, Taylor et al is of independent arithmetic and popular interest because it can be used to prove that Fermat's Last Theorem is true; that is, if we have

$$a^n + b^n = c^n, \text{ with } a, b, c \in \mathbf{Z} \text{ and } 3 \leq n \in \mathbf{N}, \tag{5.34}$$

then $abc = 0$. The basic idea of the proof is as follows: let (a, b, c) be a nontrivial solution to the Fermat equation (so $abc \neq 0$), and define an elliptic curve $E_{a,b,c}$ by

$$E : y^2 = x(x - a^n)(x + b^n).$$

(We note that if we know a and b, then we know c, so this equation does not need to include c).

This is called a *Frey curve* or a *Frey-Hellegouarch curve*; it was introduced in [Hellegouarch (1974/75)] and in [Frey (1986)]. It can be shown using level-lowering results of Ribet and others that if it exists then this curve is modular, so has a nonzero modular form f_E associated to it. By level-lowering results, it can be shown that the level of this modular form can be taken to be 2. However, it can also be shown using the dimension formulae that $S_2(\Gamma_0(2)) = 0$, so there is no such modular form, and hence our assumption that we have a nontrivial solution to (5.34) must be false.

A proof of enough of the Modularity Conjecture to prove Fermat's Last Theorem can be found in the book [Cornell *et al.* (1997)]; Chapter 1 gives a cogent summary of the proof. A briefer overview of the proof of the

Modularity Conjecture for a general audience can be found in [Darmon (1999)].

The Ramanujan-Petersson Conjecture can also be related to the Hasse bound for elliptic curves, which bounds the number of points on an elliptic curve $E(\mathbf{F}_p)$; the following theorem for elliptic curves was proved in the 1930s by Hasse (although *not* with the proof given below!); see Theorem 1.1 of Chapter V of [Silverman (1992)] for a more orthodox proof.

Theorem 5.28 (Hasse). *Let E be an elliptic curve and let $\mathbf{F}$ be a finite field with p elements. Then*

$$|\#E(\mathbf{F}) - p - 1| \leq 2\sqrt{p}.$$

Proof. Using the theorem that all elliptic curves over $\mathbf{Q}$ are modular, we see that $a_p \leq 2\sqrt{p}$, and hence that Hasse's theorem holds. □

It is interesting to see that the two bounds on coefficients of cuspidal modular eigenforms of weight 2 that we have seen are exactly the same. Moreover, it can be seen by explicit computation of examples that they are optimal, in the sense that there are elliptic curves which achieve the natural generalization of the Hasse bound as we have stated it to fields with p^n elements.

Another example of where modular forms and elliptic curves are linked will be seen in Section 5.9, where the concept of elliptic curves with complex multiplication will be introduced. We will see how these can be linked to an interesting class of modular forms.

5.8 Theta functions and their applications

In Chapter 1, we mentioned that one historical application of modular forms was to bound the number of ways that one can write a positive integer as a sum of squares. We also recall that some of the earliest modular forms to be written down were theta functions, in the work of Bernoulli.

We will now consider the theory of these objects. A good reference for this is [Miyake (2006)], Section 4.9; [Serre (1973a)], Section 6.5 also considers theta functions, as does the older book [Ogg (1969)], Chapter VI, and [Koblitz (1993)], Section III.4 considers the specific theta function of weight 1 and level $\Gamma_0(4)$

$$g(z) := \sum_{x,y \in \mathbf{Z}} q^{x^2+y^2}$$

in detail, proving that it is a modular form by considering how it transforms under the action of a set of generators for $\Gamma_0(4)$.

There is also a brief discussion of theta functions in a wider context in [Hida (2006)], which gives formulae for the number of representations of n as a sum of m squares, where m runs from 2 to 8. This is still a subject of research in modern times; the paper [Shimura (2002)] considers this question in detail. The recent book [Moreno and Wagstaff (2006)] is an accessible reference for representations of integers by sums of squares, and [Grosswald (1985)] is a comprehensive history of the theory of this subject. Another classical and very good source is Dickson's comprehensive history of number theory, volume III [Dickson (1966)]; chapters X and XI of this give an overview of results on the representation of integers by quadratic forms.

We need to define what we mean by a representation of the non-negative integer n as a sum of m squares. We will say that two representations are distinct if their constituents differ in order or sign, and we allow possible multiple zeroes; for instance, there are 8 representations of 5 as a sum of two squares under our definition:

$$\begin{aligned} 5 &= 1^2 + 2^2 = (-1)^2 + 2^2 = 1^2 + (-2)^2 = (-1)^2 + (-2)^2 \\ &= 2^2 + 1^2 = (-2)^2 + 1^2 = 2^2 + (-1)^2 = (-2)^2 + (-1)^2. \end{aligned}$$

It is a classical result of number theory (conjectured and announced by Fermat, proved by Euler) that there are 8 representations of a prime number p as a sum of 2 squares if $p \equiv 1 \mod 4$, and no representations of p as a sum of 2 squares if $p \equiv 3 \mod 4$.

Similarly, we will call two representations of an integer n by a quadratic form distinct if their constituents differ in order or sign.

5.8.1 *Representations of n by a quadratic form in an* even *number of variables*

There are two distinct cases here; either we are trying to represent n with a quadratic form in an even number of variables, or we are trying to represent it by a quadratic form in an odd number of variables. It will turn out to be much easier to construct the theory in the even number case, which we will consider first.

In this section, we will take A to be a square, symmetric, integer matrix of dimension $d \geq 1$ which is positive definite. If we are considering the sum-of-n-squares problem, we will take A to be the identity matrix of dimension n.

For a non-negative integer n, we define

$$r_A(n) := \#\left\{x \in \mathbf{Z}^d : x^T A x = n\right\},$$

and define a holomorphic function on $\mathcal{H}$ by

$$f(z) := \sum_{n=0}^{\infty} r_A(n) \cdot e^{2\pi i n z}.$$

It is clear that one can compute $r_A(n)$ for any given A and n in a finite time, so we can, in theory, determine the Fourier expansion of $f(z)$ up to any given precision using a computer. More specifically, if A corresponds to a diagonal quadratic form in d variables, we can see clearly that we need to consider about $(\sqrt{n})^d$ tuples to compute $r_A(n)$; this is a reasonable amount of work if n and d are not too large.

If we choose $A = \left(\begin{smallmatrix} 1 & 0 \\ 0 & 1 \end{smallmatrix}\right)$ then we see that f and the form g we defined in the previous section have the same Fourier expansion, and in fact we see that they give the same modular form.

We will now quote a theorem from [Miyake (2006)] which will show that $f(z)$ is a modular form, and give its weight and level. Take $h \in \mathbf{Z}^d$ and a positive integer N which satisfies

$$NA^{-1} \in M_d(\mathbf{Z}) \text{ and } Ah \in N\mathbf{Z}^d.$$

We define a differential operator Δ_A (not to be confused with the Δ function, which is a modular form of weight 12) acting on polynomials in d variables by

$$\Delta_A = \sum_{i,j=1}^{d} b_{i,j} \cdot \frac{\partial^2}{\partial x_i \partial x_j}, \text{ where } A^{-1} = [b_{i,j}].$$

Let $P(x)$ be a homogenous polynomial of degree ν with complex coefficients in d variables $\{x_i\}$. We say that P is a *spherical function of degree ν with respect to the matrix A* if $\Delta_A(P(x)) = 0$.

It can be shown that every spherical function of degree ν is of the form

$$\begin{aligned} P(x) &= \text{a constant, if } \nu = 0, \\ &= q^T A x, \text{ where } q \in \mathbf{C}^d, \text{ if } \nu = 1, \\ &= \text{a linear combination of } (q_i^T A x)^\nu,\ q_i \in \mathbf{C}^d,\ q_i^T A q_i = 0, \text{ if } \nu > 1. \end{aligned}$$

They are therefore explicitly computable objects.

We now define *theta functions* (which are also called *theta series* in the literature). With the notation that we have just defined, if $z \in \mathcal{H}$ we define

$$\theta(z; h, A, N, P) := \sum_{\substack{m \equiv h \mod N \\ m \in \mathbf{Z}^d}} P(m) \cdot e\left(\frac{m^T A m}{2N^2} \cdot z\right) \tag{5.35}$$

$$= \sum_{n=0}^{\infty} a(n, h, A, N, P) \cdot e\left(\frac{n}{2N^2} \cdot z\right); \tag{5.36}$$

where

$$a(n, h, A, N, P) = \sum_{\substack{m^T A m = n \\ m \equiv h \mod N}} P(m).$$

We obtain the series (5.36) by collecting the coefficients of each power of $e(z/2N^2)$. Doing this allows us to look at the Fourier series of θ, which is what we are interested in.

We now quote the following result, which appears in [Miyake (2006)] as Corollary 4.9.4.

Theorem 5.29. *Let A be a square symmetric positive definite matrix of even dimension d, let P be a spherical function of degree ν with respect to A and let $h \in \mathbf{Z}^d$. Then*

$$\theta(2z; h, A, N, P) \in M_k(\Gamma_1(4N)),$$

where $k = d/2 + \nu$. Furthermore, if $\nu \geq 1$, then $\theta(2z; h, A, N, P)$ is a cusp form.

Proof. There are two parts to this proof; one has to show that θ is holomorphic on $\mathcal{H}$ and at the cusps of $\Gamma_1(4N)$, and one has to show that it satisfies the correct transformation formula. The second part is long and technical; we refer to Section 4.9.1 of [Miyake (2006)], especially Lemma 4.9.1, Lemma 4.9.2 and Theorem 4.9.3, or Section III.4 of [Koblitz (1993)], Lemmas 1–4.

We will now show that $\theta := \theta(2z; h, A, N, P)$ is holomorphic on $\mathcal{H}$. We recall that the matrix A is positive definite, so there exist positive real numbers L and U such that

$$L \cdot x^t x \leq x^t A x \leq U \cdot x^t x,$$

for every $x \in \mathbf{R}^d$. The Cauchy-Schwarz inequality tells us that there exists a positive real number T such that

$$|P(x)| \leq (T x^t x)^{\nu/2},$$

for all $x \in \mathbf{R}^d$. Hence we have the following bound on the Fourier coefficients of θ:

$$|a(n,h,A,N,P)| = \sum_{\substack{m^T Am = n \\ m \equiv h \mod N}} |P(m)| = O(n^{(d+\nu)/2}).$$

We now quote the following lemma, which appears in [Miyake (2006)] as Lemma 4.3.3:

Lemma 5.30. *Let $\{a_i\}_{i=1}^{\infty}$ be a sequence of complex numbers, and define the following Fourier series:*

$$f(z) := \sum_{n=1}^{\infty} a_n q^n, \text{ with } q := e^{2\pi i z}.$$

If $a_n = O(n^v)$ for some $v > 0$, then $f(z)$ converges absolutely and uniformly on any compact subset of $\mathcal{H}$, and $f(z)$ is holomorphic on $\mathcal{H}$.

Using this lemma, we see that θ is holomorphic on $\mathcal{H}$ as required. It is clearly holomorphic at ∞ because it has no nonzero Fourier coefficients a_n with $n < 0$ (because A is positive definite). In a similar way, we can show that it is holomorphic at all of the other cusps, by choosing suitable $\left(\begin{smallmatrix} a & b \\ c & d \end{smallmatrix}\right)$ and showing that $\theta|[\left(\begin{smallmatrix} a & b \\ c & d \end{smallmatrix}\right)]_k$ has a Fourier series with no terms with negative exponents, which will show that it has no poles at those cusps.

The proof that θ transforms correctly is long and technical, and can be found in pages 187–192 of [Miyake (2006)]. □

We can now estimate the number of ways that a positive integer n can be written as a sum of $2r$ squares by forming a theta function $\theta(z)$ using the identity matrix of dimension $2r$, and then using the Ramanujan-Petersson conjecture to find bounds for the coefficients; we note that θ is a sum of modular forms which involves an Eisenstein series, because it has nonzero constant term. This leads to the following result:

Corollary 5.31. *Let $2k$ be a positive integer. Then the number of ways that a positive integer n can be represented as a sum of $2k$ squares is $O(n^{k-1})$.*

Proof.

We form the theta function $f(z) := \theta(2z; 0, I_k, 1, 1)$ where I_{2k} is the $2k \times 2k$ identity matrix, and the 1 is the constant function, which we have seen is a spherical function of degree 0. We see that the nth Fourier coefficient of f is precisely the number of ways that n can be written as a sum of $2k$

squares. Using Theorem 5.29, we see that $f(q)$ is the Fourier expansion of a modular form f of integer weight k for $\Gamma_1(4)$, which is not a cusp form.

We now consider the dimension of the Eisenstein subspace of $M_k(\Gamma_1(4))$. By the dimension formulae given in Chapter 3, we see that this is 1 if $k = 1$, 2 if k is odd or $k = 2$, and 3 otherwise.

We will now write down a basis for the space of Eisenstein series in each of these cases and check the order of magnitude of the Fourier coefficients. If $k = 1$, we can take $\{\sum_{x,y\in\mathbf{Z}} q^{x^2+y^2}\}$, if $k = 2$, we take $\{E_2(q) - 2E_2(q), E_2(q^2) - 2E_2(q^4)\}$, and if $k \geq 4$ is even we take $\{E_k(q), E_k(q^2), E_k(q^4)\}$. If $k \geq 3$ is *odd*, then we use Theorem 2.21 to construct two distinct Eisenstein series.

We can show directly that all of these Eisenstein series have Fourier coefficients which are $O(n^{k-1})$ (this is clear for all of them except the odd weight forms, for which we refer to Theorem 2.23). We now write f as a linear combination of the Eisenstein series and a cusp form g, and we then apply the Ramanujan-Petersson conjecture (Theorem 5.21) to g to obtain the bound for the coefficients of f. □

Remark 5.32. We note that a similar proof shows us that the number of representations of a positive integer n by a positive definite quadratic form in $2k$ variables is $O(n^{k-1})$.

We noted in the course of the proof that the number of Eisenstein series for $k \geq 3$ is 3 if k is even and 2 if k is odd; we will now explain this discrepancy in a more concrete way. In the even case, we have one Eisenstein series for each of the three cusps $(0, -1/2, \infty)$, but the structure theorem for this space of modular forms tells us that we can write odd weight forms as $\theta^2 \cdot g$, where g has even weight. It can be shown that θ^2 has a unique zero at $-1/2$, so therefore we only need to take the sum of three linearly independent modular forms to obtain a cusp form, so the Eisenstein subspace has dimension 2 instead of dimension 3.

We can actually do more than just get a bound for the number of representations of an integer as the sum of $2k$ squares, if $2k$ is small. This was known by Glaisher at the beginning of the 20th century; he proves increasingly complicated formulae for the number of representations of a positive integer as a sum of larger and larger numbers of squares in [Glaisher (1907a,b,c)], which are all in a single volume of the *Quarterly Journal of Pure and Applied Mathematics*. If $k \leq 8$ then the space $S_k(\Gamma_1(4))$ has dimension 0, so we can write the theta function in terms of the Eisenstein series which form a basis for $M_k(\Gamma_1(4))$.

The problem that one encounters as k grows is that eventually the space $S_k(\Gamma_1(4))$ becomes nonzero, and while the Eisenstein series have coefficients which are well-known arithmetical variations on the sum of divisors function, the coefficients of cusp forms are more mysterious and cannot be written down so cleanly. However, they *are* smaller than the coefficients of Eisenstein series (by the Ramanujan-Petersson conjecture, or by the weaker result that we proved as Theorem 5.18), so we can obtain order-of-magnitude formulae such as the one in Corollary 5.31.

We follow Koblitz in performing this calculation for 4 squares, and in Exercise 6, we perform a similar calculation for 8 squares. We note that there is a proof in [Hardy and Wright (1979)] of this theorem which uses formulae for trigonometric series, amongst other things (Theorem 386, in Chapter XX, Section 12).

Theorem 5.33. *Let n be a non-negative integer. Then we can write n in $r_4(n)$ ways as a sum of four squares, where*

$$\begin{aligned} r_4(n) &= 8\sigma_1(n), \quad \text{if } n \text{ is odd}, && (5.37)\\ &= 24\sigma_1(n') \text{ where } n = 2^a n' \text{ is even, with } n' \text{ odd}. && (5.38) \end{aligned}$$

Proof. We define Θ to be the series

$$\Theta(q) := \sum_{x,y,z,w \in \mathbf{Z}} q^{x^2+y^2+z^2+w^2} \in M_2(\Gamma_0(4)).$$

We will now find a basis for $M_2(\Gamma_0(4))$ and consider the action of the Hecke operators on this basis. By using the dimension formula, we see that $\dim M_2(\Gamma_0(4)) = 2$. We see that Θ is an element of $M_2(\Gamma_0(4))$, and that $D(q) := E_2(q) - 2E_2(q^2) \in M_2(\Gamma_0(2))$ is another (see Exercise 9 for a proof that this is a modular form). We will also consider $D(q^2)$, which is an element of $M_2(\Gamma_0(4))$ as an oldform coming from $M_2(\Gamma_0(2))$; we will find an equality relating the Fourier coefficients of Θ, $D(q)$ and $D(q^2)$.

The Fourier expansions of these modular forms begin

$$\begin{aligned} \Theta(q) &= 1 + 8q + 24q^2 + 32q^3 + 24q^4 + 48q^5 + 96q^6 + 64q^7 + O(q^8),\\ D(q) &= 1 + 24q + 24q^2 + 96q^3 + 24q^4 + 144q^5 + 96q^6 + 192q^7 + O(q^8),\\ D(q^2) &= 1 + 24q^2 + 24q^4 + 96q^6 + O(q^8). \end{aligned}$$

As these three forms are pairwise linearly independent, we see that any pair of them form a basis. In particular, we see that

$$\Theta(q) = \frac{D(q) + 2D(q^2)}{3} = \frac{E_2(q) - 4E_2(q^4)}{3}.$$

If n is odd, then this is enough to prove the theorem, because $E_2(q^4)$ does not contribute to the q^n term of the Fourier expansion, and the nth Fourier coefficient of $E_2(q)/3$ is $8\sigma_1(n)$, as required.

We now assume that n is even. We will consider the cases where $n = 2m$ (m odd) and $n = 4m$ (m arbitrary) separately.

If $n = 2m$ with m odd then $4E_2(q^4)/3$ does not contribute to the q^n term of the Fourier equation, so we have only to consider $E_2(q)/3$. We have the following identity for its Fourier coefficient:

$$8\sigma_1(n) = 8 \sum_{1 \leq d | n} d = 8 \sum_{\substack{d|n \\ d \text{ odd}}} d + 8 \sum_{\substack{2d|n \\ d \text{ odd}}} 2d = 24 \sum_{\substack{d|n \\ d \text{ odd}}} d = 24\sigma_1(m).$$

This is exactly the formula given in the theorem. We now generalize this argument to the case where $n = 4m$. In this case, the $4E_2(q^4)/3$ term does contribute, so we must take it into consideration.

The q^n term in the Fourier expansion of $\Theta(q)$ is given by

$$r_4(n) = 8\sigma_1(4m) - 8 \cdot 4\sigma_1(m).$$

Now, we can rewrite the sum-of-divisors function in the following way; let $n = 2^a n'$, with n' odd (so $m = 2^{a-2}n'$):

$$\sigma_1(n) = \sum_{1 \leq d | n} d = \sum_{i=0}^{a} 2^i \cdot \sum_{1 \leq d | n'} d = (2^a - 1)\sigma_1(n').$$

We can now compute $r_4(n)$ in this final case using this formula:

$$\begin{aligned} r_4(n) &= 8\sigma_1(4m) - 8 \cdot 4\sigma_1(m) \\ &= 8(2^{a+2} - 1)\sigma_1(n') - 32(2^a - 1)\sigma_1(n') \\ &= 24\sigma_1(n'), \end{aligned}$$

as required. □

We notice that in particular we have that $r_4(n) > 0$; this means that as a by-product of our proof we have proved Legendre's theorem that every integer is representable as the sum of four squares.

There is an alternative proof, given as Exercise III.5.2 of [Koblitz (1993)], which uses the theory of Hecke operators; one finds a basis of eigenforms for T_2 and then analyzes the action of T_n to show that the coefficients of Θ are given by the formulae in the theorem. This uses the key fact that a basis of eigenforms for T_2 is in fact a basis of eigenforms for every T_p.

Finally, we will note the following result on the space of modular forms for $\Gamma_1(4)$.

Proposition 5.34. *Let k be a positive integer, and define $F(q) = \sum_{n>0 \text{ odd}} \sigma_1(n)q^n$ and $\theta(q) = \sum_{n=-\infty}^{\infty} q^{n^2}$. Assign weight 2 to F and weight 1 to θ^2. The space of modular forms for $M_k(\Gamma_1(4))$ is given by polynomials which are the sum of monomials of weight k in F and θ^2. Moreover, the space $S_k(\Gamma_1(4))$ is given by the graded ideal of polynomials of weight k which are divisible by $\theta^4 F(\theta^4 - 16F)$.*

We note that F can be viewed either as the η-quotient $\eta(q^4)^8/\eta(q^2)^4$ or as the linear combination of Eisenstein series $-1/24(E_2(q) - 3E_2(q^2) + 2E_2(q^4))$; this can be proved in the standard way, by computing the dimension of $M_2(\Gamma_0(4))$, finding the Sturm bound and then checking explicitly that the Fourier expansions are the same.

Proof. To prove this, we use the dimension formulae from Chapter 3 to show that

$$\dim S_k(\Gamma_1(4)) = \dim M_{k+6}(\Gamma_1(4)), \tag{5.39}$$

that $S_5(\Gamma_1(4)) = \mathbf{C}\theta^2 F(\theta^4 - 16F)$ and $S_6(\Gamma_1(4)) = \mathbf{C}\theta^4 F(\theta^4 - 16F)$, and that the spaces of modular forms for $\Gamma_1(4)$ of weights between 0 and 5 are all generated by polynomials in F and θ^2 of the correct weights. One then proves that there is an isomorphism between cusp forms for $\Gamma_1(4)$ of weight $k+6$ and modular forms for $\Gamma_1(4)$ of weight k, for every non-negative integer k.

We leave the computational details to the reader as Exercise 15. □

This proposition tells us, amongst other things, that we can write theta functions associated to the problem of representing n as a sum of an even number of squares in terms of the functions F and θ^4.

5.8.2 *Representations of n by a quadratic form in an* odd *number of variables*

It is natural to ask "what happens if ν is odd?" Theta functions with ν odd correspond to theta functions obtained by considering the number of solutions to quadratic forms with an odd number of variables.

There is a theory of these, which goes back at least to work of Hardy in [Hardy (1920)], but they are not modular forms of the type that we have been considering so far; they are instead "half-integral weight modular forms". The reason that they are not classical modular forms is that they would have to have a fractional power of $cz + d$ in the transformation formula, and one has to be careful about the choice of the branch of the square root.

We note that MAGMA now supports computations with half-integral weight modular forms, using the command `HalfIntegralWeightForms`.

A good and accessible reference for these is Chapter IV of [Koblitz (1993)], where we have the following theorem, similar to Proposition 5.34:

Theorem 5.35 (The space of half-integral modular forms). *Let k be a positive odd integer. Let $\theta(z) = \sum_{n=-\infty}^{\infty} q^{n^2}$ and let $F(z) = \sum_{n>0 \text{ odd}} \sigma_1(n) q^n$. Assign weight $1/2$ to θ and weight 2 to F. Then the space $M_{k/2}(\Gamma_0(4))$ is the space of all polynomials in the graded ring $\mathbf{C}[\theta, F]$ which are sums of monomials each with weight $k/2$; we note that some authors write this as $M_{k/2}(\widetilde{\Gamma}_0(4))$.*

For instance, if $k = 1$ then the space of forms has dimension 1; it is generated by θ. There is a notion of cusp forms of weight $k/2$ in this setting; these are the modular forms of weight $k/2$ in the ideal generated by $\theta F(16F - \theta^4)$ within $\mathbf{C}[\theta, F]$, and are written either as $S_{k/2}(\Gamma_0(4))$ or $S_{k/2}(\widetilde{\Gamma}_0(4))$; this corresponds to the cusp forms for $\mathrm{SL}_2(\mathbf{Z})$ being those modular forms which are in the ideal generated by Δ. The cusp form of lowest weight here is $\theta F(16F - \theta^4)$, which has weight $9/2$. We also note that θ and F are linearly independent over $\mathbf{C}$.

We can prove the following result on the dimension of the space of modular forms of half-integral weight and level $\Gamma_0(4)$:

Proposition 5.36. *Let k be an odd positive integer. Then*

$$\dim M_{k/2}(\Gamma_0(4)) = 1 + \left\lfloor \frac{k}{4} \right\rfloor,$$

$$\dim S_{k/2}(\Gamma_0(4)) = \max\left(1 + \left\lfloor \frac{k-8}{4} \right\rfloor, 0\right).$$

Proof. If $k = 1$ or $k = 3$ then there is only one (non-cuspidal) modular form of weight $k/2$ (θ or θ^3, respectively). In these cases we see that the formulae hold.

If $k \geq 5$ then there are $1 + \lfloor k/4 \rfloor$ powers of F which have pure weight less than $k/2$ (the 1 comes from the trivial power of F, F^0). Each of these contributes one monomial, and every modular form of half-integral weight is a polynomial in these monomials.

Similarly, from the discussion above we see that every cusp form is a polynomial of weight $k/2$ which is divisible by a certain modular form of weight $9/2$, so in effect we are looking at modular forms of weight $(k-9)/2$, which gives the formula in the proposition. □

We note that we have seen the same form of argument about the dimension of the space of cusp forms before, when we were proving dimension formulae for $S_k(\mathrm{SL}_2(\mathbf{Z}))$ in Chapter 3.

Finding the number of representations of an integer as a sum of three squares is much more difficult than finding the the number of representations in terms of either two or four squares; we recall that there are formulae which, given representations of two integers m and n in terms of two or four squares, give a representation of mn in terms of two or four squares. There is no such formula for sums of three squares. The following theorem, from [Armitage and Eberlein (2006)], does give an arithmetic interpretation of the coefficients of the modular form θ^3 (which has weight 3/2). We let $ax^2 + 2bxy + cy^2$ be a binary quadratic form, with discriminant[9] $D := b^2 - ac$. We say that two such forms are *equivalent* if one can be transformed into the other by a substitution of the form $(x, y) \mapsto (ax + by, cx + dy)$, where $\left(\begin{smallmatrix} a & b \\ c & d \end{smallmatrix}\right) \in \mathrm{SL}_2(\mathbf{Z})$. We say that two forms which are equivalent under this action of $\mathrm{SL}_2(\mathbf{Z})$ are *in the same class.*

It is a sign of how different areas of mathematics are related to one another that the fundamental domain shown in Figure 2.1 is also a fundamental domain for binary quadratic forms, in that every positive definite binary quadratic form can be shown to be equivalent to a *reduced quadratic form* (which is defined to be one where either $-a < b \le a \le c$ or $0 \le b \le a = c$), and one of the complex roots of a reduced quadratic form will lie within the standard fundamental domain F.

We quote the following theorem (which appears as Theorem 11.1 of [Armitage and Eberlein (2006)]) without proof.

Theorem 5.37. *Let n be a positive integer, and let $G(n)$ be the number of classes of binary forms of discriminant $-n$, and $F(n)$ be the number of classes of such forms where at least one of a and c is odd. Then we have that*

$$r_3(n) = 24F(n) - 12G(n).$$

[9]The notation Δ is sometimes used for the discriminant, but it already has too many definitions in the theory of modular forms.

Proof. The proof is given as Chapter 11 of [Armitage and Eberlein (2006)]; first one shows that the Fourier coefficients of θ^3 give $r_3(n)$, and then one uses the theory of elliptic functions[10] to prove that $r_3(n)$ is given by the difference of classes. □

One new feature of modular forms of half-integral weight is that the Hecke operators T_n are identically zero if the positive integer n is not a perfect square (see Proposition IV.3.12 of [Koblitz (1993)]). If n is a perfect square, then we have the following result:

Theorem 5.38. *Let N be a positive integer such that $4|N$, let $\chi : (\mathbf{Z}/N\mathbf{Z})^\times \to \mathbf{C}^\times$ be a Dirichlet character, let p be a prime and let $k = 2l+1$ be a positive odd integer. Let $f(z) = \sum_{n=0}^{\infty} a_n q^n \in M_{k/2}(\Gamma_0(N), \chi)$. Then we have*

$$T_{p^2} f(z) = \sum_{n=0}^{\infty} b_n q^n,$$

where

$$b_n = a_{p^2 n} + \chi(p) \cdot \left(\frac{(-1)^l n}{p}\right) p^{l-1} a_n + \chi(p^2) p^{k-2} a_{n/p^2};$$

(we have the convention that $a_{n/p^2} = 0$ if $p^2 \nmid n$).

We note that if $p \mid N$ then $\chi(p) = \chi(p^2) = 0$, so then we have the much simpler formula $b_n = a_{p^2 n}$.

The proof of this theorem involves considering the action of the Hecke operators explicitly on a modular form of half-integral weight, and then splitting it and rewriting each part in order to give the three summands in the formula. We will not give this here; it can be found in [Koblitz (1993)] as Proposition IV.3.13.

As in the integer weight case, one can find a basis of modular forms for $M_{k/2}(\Gamma_0(N), \chi)$ which are eigenforms for all of the T_{p^2} for $p \nmid N$, and in many circumstances we can find a basis for *all* of the Hecke operators.

5.8.3 *The Shimura correspondence*

In [Shimura (1973)], the following theorem relating forms of half-integral weight $k/2$ to forms of integral weight $k-1$ is given:

[10] An *elliptic function* f is a meromorphic complex function such that there exists a lattice L in $\mathbf{C}$ such that $f(z+\omega) = f(z)$ for every $z \in \mathbf{C}$ and $\omega \in L$. [Armitage and Eberlein (2006)] and [Whittaker and Watson (1996)], Chapter XX are good references for these; [Silverman (1992)], Section VI.2 gives an account which relates them to the theory of elliptic curves.

Theorem 5.39 ([Shimura (1973)], Main Theorem). *Let $k \geq 3$ be an odd integer, let N be a positive integer, let χ be a Dirichlet character modulo $4N$, and let $f \in S_{k/2}(\Gamma_0(4N), \chi)$ be a cusp form with Fourier expansion $f(q) = \sum_{n=1}^{\infty} a_n q^n$ which is an eigenform for all the T_{p^2} with eigenvalues λ_{p^2}.*

Define a function $g(q) = \sum_{n=1}^{\infty} b_n q^n$ by

$$\sum_{n=1}^{\infty} b_n n^{-s} = \prod_{p \text{ prime}} \frac{1}{1 - \lambda_{p^2} p^{-s} + \chi(p)^2 \cdot p^{k-2-2s}}.$$

Then $g(q)$ is the Fourier expansion of a modular form in $M_{k-1}(\Gamma_0(2N), \chi)$, and if $k \geq 5$, g is a cusp form.

Another reference for this section is [Koblitz (1993)], Section IV.4; the version of the theorem given here reflects and follows Koblitz's presentation.

Let us consider a numerical example, similar to that given in [Koblitz (1993)]. Using the dimension formulae given in Proposition 5.36, we see that the space of cuspforms for $\Gamma_0(4)$ of weight $11/2$ is one-dimensional, and is spanned by $f := \theta^3 F(16F - \theta^4)$. Using the dimension formulae for integer-weight modular forms, we see that the space in which g lies, $S_{10}(\Gamma_0(2))$, is one-dimensional, so g is a constant multiple of the unique normalized cusp form h of that weight and level.

Because a modular form for $\Gamma_0(2)$ is also a modular form for $\Gamma_0(4)$, we can write modular forms for $\Gamma_0(2)$ as polynomials in F and θ. We therefore need to be able to find the Fourier expansion of h. We will use the fact that there are modular forms in $S_{10}(\Gamma_0(4))$ with Fourier expansions $h(q)$ and $h(q^2)$; we will see in Exercise 16 that this uniquely identifies h; in fact, we can show that

$$h(q) = q + 16q^2 - 156q^3 + 256q^4 + 870q^5 + O(q^6). \tag{5.40}$$

Using Proposition 5.34, we can therefore write $h(q)$ in terms of the Fourier expansions of the modular forms $F(q)$ and $\theta^2(q)$, and we find that $g(q)$ is a constant multiple of

$$h(q) = \sum_{n=1}^{\infty} b_n q^n = \theta^4 F(\theta^4 - 16F)(\theta^8 - 256F^2);$$

It can be shown (we refer to Proposition 14 of Chapter IV of Koblitz for the details) that the following relationship holds between the a_{p^2} and the b_p:

$$b_2 = a_4; \; b_p = \lambda_{p^2} = a_{p^2} + \left(\frac{-1}{p}\right) p^4 \text{ if } p \geq 3.$$

This is an example of what Koblitz calls a "remarkable numerical identity"; it relates coefficients of one sum of η-quotients to (different) coefficients of another sum of η-quotients; the bare statement without the context of the modular forms interpretation seems, in Koblitz's words, to be "rather outlandish". These can be compared to the identities exhibited in Section 5.5.

5.9 CM modular forms

We can use the results of the previous section to find modular forms which have "many" zero Fourier coefficients, in some rigorous sense. Recall that Lehmer's Conjecture says that $\tau(n) \neq 0$ for all $n \geq 1$; it is believed that *none* of its Fourier coefficients are zero.

However, if we consider the modular form $\theta^2 \in M_1(\Gamma_1(4))$, we see that at least a quarter of its Fourier coefficients are zero, because $a^2 + b^2 \not\equiv 3 \mod 4$ if a and b are integers. Moreover, if we consider the coefficients a_p where p is an odd prime — this is reasonable because θ^2 is an eigenform — we see that in the limit as $p \to \infty$ half of these coefficients are zero.

Similarly, if $f(x) = x^2 + Ny^2$ is a positive definite quadratic form, we see that there will be infinitely many positive integers which are *not* represented by it, so the theta function defined by this form will be a modular form of weight 1 for $\Gamma_1(4N)$ with infinitely many zero Fourier coefficients.

This is not the only class of examples; the space $S_5(\Gamma_1(4))$ is 1-dimensional, and is spanned by

$$f(q) = \eta(q)^4\eta(q^2)^2\eta(q^4)^4 = q\prod_{n=1}^{\infty}\left((1-q^n)^2(1-q^{2n})(1-q^{4n})^2\right)^2;$$

we see by considering the right-hand side of this η-product that if $n \equiv 3 \mod 4$ then the nth Fourier coefficient of f is 0, so there are infinitely many zero Fourier coefficients.

We introduce the following definition from [Ribet (1977)], to formalize our intuition as to what "many" Fourier coefficients being zero means:

Definition 5.40. Let ε be a nontrivial Dirichlet character and let $f \in M_k(\Gamma_0(N), \chi)$ be a modular new eigenform. We say that f *has complex multiplication by* ε if

$$\varepsilon(p)a_p = a_p \tag{5.41}$$

for all primes p in a set of density 1. If the kernel of ε in $\mathrm{Gal}(\overline{\mathbf{Q}}/\mathbf{Q})$ corresponds to a field F then we say that f *has complex multiplication* *by F*. Also, if a modular form f has complex multiplication, we often say that *f is a CM modular form* or *f is a CM form*.

Remark 5.41. We see that ε here has to be a quadratic character, because Theorem 4.29 tells us that the modular form $f \otimes \varepsilon$ is a modular form for $\Gamma_1(ND^2)$, where D is the conductor of ε. If 5.41 holds, then we know that $f = f \otimes \varepsilon$, so we have that $\chi\varepsilon^2 = \chi$, so $\varepsilon^2 = 1$.

We see that the modular form f we defined above has complex multiplication by the nontrivial Dirichlet character of conductor 4; we can take the set of primes for the definition to be those primes congruent to 3 modulo 4. We will show that this works for infinitely many $S_k(\Gamma_1(4))$ in Exercise 20.

Another example is the unique normalized cuspform $g \in S_2(\Gamma_0(32))$; this is given by

$$g(q) = \eta(q^4)^2\eta(q^8)^2 = q\prod_{n=1}^{\infty}\left((1-q^{4n})(1-q^{8n})\right)^2.$$

Again, we see that we can choose a Dirichlet character (modulo 32) such that (5.41) holds. We leave the details as Exercise 17.

The arithmetic interest in these forms does not come solely from the fact that they have "many" zero Fourier coefficients. Let E be an elliptic curve over $\mathbf{Q}$. There are many isogenies from E to itself; for instance, there are the multiplication-by-m maps, where $m \in \mathbf{Z}$. If the endomorphism ring of E is strictly greater than $\mathbf{Z}$, then we say that *the elliptic curve E has complex multiplication*. For example, let us consider

$$E : y^2 = x^3 - x.$$

We see that the endomorphism ring of E also contains the element

$$[i] : (x, y) \mapsto (-x, iy);$$

where i is a square root of -1. We see that this is an endomorphism of $E/\mathbf{Q}$ which is *not* multiplication by a rational integer, and therefore that $E/\mathbf{Q}$ has complex multiplication (by a ring containing $\mathbf{Z}[i]$, which is in fact equal to $\mathbf{Z}[i]$).

CM elliptic curves over $\mathbf{Q}$ are very interesting; their extra structure means that many important conjectures were first proved for CM curves and only later for general elliptic curves. A good example of this is the Shimura-Taniyama-Weil conjecture, which was proved for CM curves in the 1970s by Shimura [Shimura (1971)], long before the more general proofs of Wiles and others in the 1990s.

There is a literature on complex multiplication which includes [Shimura (1998)]. The concept of complex multiplication can be generalized to higher-dimensional abelian varieties also; we will not consider this here.

5.10 Lacunary modular forms

Following on from the previous section, we now briefly consider the notion of lacunary modular forms, as defined in [Serre (1985)]. We will see that these have a close relationship to CM forms.

Definition 5.42. Let $f(q) = \sum_{n=0}^{\infty} a_n q^n$ be a power series. Define $M_f(x)$ to be the number of $n \leq x$ such that $a_n \neq 0$. We say that f is *lacunary* if

$$\lim_{x\to\infty} \frac{M_f(x)}{x} = 0;$$

in other words, the density of the nonzero coefficients is zero.

If f is a modular form with Fourier expansion $f(q)$, then we say that f *is a lacunary modular form.*

We now state and prove a famous theorem from classical number theory.

Theorem 5.43 (Euler's pentagonal number theorem). *The following identity of formal power series holds:*

$$\prod_{n=1}^{\infty}(1-q^n) = \sum_{k=-\infty}^{\infty}(-1)^k q^{k(3k-1)/2}. \tag{5.42}$$

We note that this theorem as stated does not involve modular forms, but we see that the left hand side of (5.42) is $\eta(q)/q^{1/24}$, which can be thought of as having weight $1/2$, while the right hand side resembles a theta function in one variable, which again we can think of as having weight $1/2$.

Proof. Firstly, we note that the inverse of the left hand side of (5.42) is given by

$$\prod_{k=1}^{\infty}(1-q^k)^{-1} = \sum_{n=0}^{\infty} p(n)q^n,$$

where $p(n)$ is the number of partitions of the integer n.[11] This means that we have

$$\left(\sum_{n=0}^{\infty} p(n)q^n\right)\cdot\left(\sum_{n=0}^{\infty} a_n q^n\right),$$

[11] There is a large literature on the properties of $p(n)$, which lies beyond the scope of this book. For an overview and introduction to the subject, see [Andrews (1976)].

where the a_n are the coefficients of the series in (5.42). By considering the nth coefficient of this, we obtain the relation

$$\sum_{i=0}^{n} a_i \cdot p(n-i) = 0, \text{ for } n \geq 0. \tag{5.43}$$

The theorem that we are to prove can be restated as

$$\begin{aligned} a_n &= 1, \text{ if } n = \frac{3k^2 \pm k}{2} \text{ and } k \text{ is even}, \\ &= -1, \text{ if } n = \frac{3k^2 \pm k}{2} \text{ and } k \text{ is odd}, \\ &= 0, \text{ otherwise.} \end{aligned}$$

We will therefore substitute these values into (5.43), and prove that the equality holds, which will prove the theorem. If we define $b_n := (3i^2 + i)/2$, then this gives us

$$\sum_{i} (-1)^i \cdot p(n - b_i) = 0,$$

where $n \geq 0$, and the summation is over all integers i such that $b_i \leq n$. We can split the summation as follows:

$$\sum_{i \text{ odd}} p(n - b_i) = \sum_{i \text{ even}} p(n - b_i) \tag{5.44}$$

We now define $\mathcal{P}(n)$ to be the set of partitions of n; we can rewrite the identity (5.44) as

$$\left| \bigcup_{i \text{ odd}} \mathcal{P}(n - b_i) \right| = \left| \bigcup_{i \text{ even}} \mathcal{P}(n - b_i) \right| \tag{5.45}$$

We leave the proof of this as Exercise 24. $\square$

The name of this theorem comes from the fact that the indices of the nonzero terms in the Fourier expansion are the generalized pentagonal numbers; the pentagonal numbers are those positive integers which are of the form $i(3i-1)/2$ for some positive integer i; they can be arranged in concentric pentagons, as we can see in Figure 5.2. The generalized pentagonal numbers are those positive integers which are of the form $i(3i-1)/2$ for an integer i; those for which i is negative do not have a figurate representation. A translation of Euler's original proof into modern language can be found in the expository article [Andrews (1983)]. There are other modern proofs

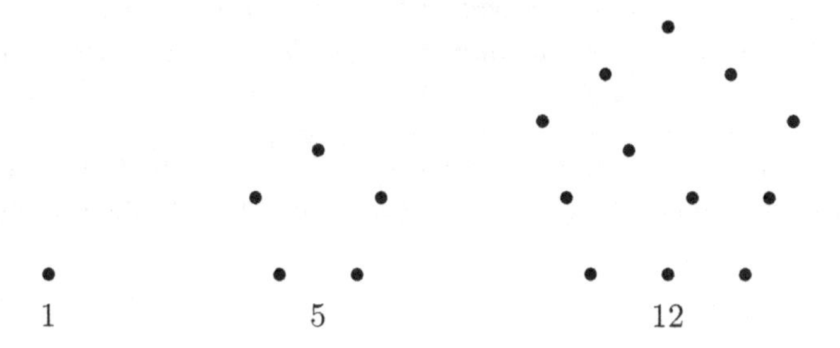

Fig. 5.2 The first three pentagonal numbers

of this theorem; one can prove it in a completely combinatorial way, or as a direct consequence of Jacobi's triple product identity:

$$1+\sum_{n=1}^{\infty}(z^n+z^{-n})x^{n^2} = \prod_{m=1}^{\infty}(1-x^{2m})(1+zx^{2m-1})(1+z^{-1}x^{2m-1}); \quad (5.46)$$

see Exercise 21.

We recall that in Chapter 1 we saw that MacMahon proved that $p(200) = 397299029388$. This was not computed by enumeration, but using the following recurrence relation on the $p(i)$, which we will quote without proof:

$$p(n) = \sum_{i=-\infty}^{\infty} (-1)^{i-1} p_{n-q_i},$$

where q_i is the ith generalized pentagonal number.

In Exercise 18, we see that Euler's pentagonal number theorem implies that $\eta(q)$ is lacunary; in fact we know that $M_\eta(x) = O(\sqrt{x})$. At the other end of the spectrum, we have already mentioned Lehmer's conjecture that $\tau(n) \neq 0$ for any $n \in \mathbf{N}$, which would make Δ a very non-lacunary function.

We can find elementary non-examples of lacunary modular forms; for instance, the Eisenstein series E_k have no nonzero Fourier coefficients, so the limit in the definition is 1, so they are not lacunary.

In [Serre (1985)], it is shown that an even power η^r of the η function is lacunary if and only if $r \in \{2,4,6,8,10,14,26\}$. In [Gordon and Robins (1995)], the authors find 45 η-products of the form $\eta(q)^r\eta(q^2)^s$ which are lacunary, and show that there are no others. In Exercise 23, we will show that the cusp form $\eta(q^6)^4$ is lacunary.

The following theorem of Serre, generalized in [Murty (1987)], shows us that the concepts of CM and of lacunarity are closely related:

Theorem 5.44. *Let f be a modular form of integral weight with Fourier expansion $f(q)$. Then f is lacunary if and only if it is a finite linear combination of CM modular forms.*

A similar classification theorem is believed to be true for half-integral weight modular forms; some progress was made towards this by [Ono (1998)].

5.11 Exercises

(1) Choose an odd positive integer n and a square integer m, and write $T_m\theta^n$ as a polynomial in F and θ.

(2) In this question we will find an expression in terms of $\sigma_1(\lambda n)$ for the number of ways that one can represent a positive integer by the quadratic form

$$Q(x, y, z, w) = x^2 + y^2 + 2z^2 + 2w^2.$$

  (a) Find the level of the theta function

$$\Theta(q) = \sum_{x,y,z,w\in\mathbf{Z}} q^{Q(x,y,z,w)},$$

  viewed as a modular form.

  (b) Use the dimension formulae to find the dimension of this space, and show that there are no cusp forms in this space.

  (c) Find a basis of Eisenstein series for this space.

  (d) Write Θ as a combination of these Eisenstein series and equate coefficients to find the required expression.

(3) Prove Theorem 5.11.

  (a) First show that the spaces $S_k(\Gamma(N))$ given in the theorem are all 1-dimensional.

  (b) Now show that $\Delta^{1/N}$ is a modular form, by considering the action of $\Gamma(N)$ or otherwise.

(4) Fill in the details in the proof of Theorem 5.33. In particular:

  (a) Find the dimension of the space of modular forms.

  (b) Compute the Sturm bound, to verify that we have computed enough terms.

  (c) Prove that the identities of sums of divisors hold.

(5) In this exercise, we will prove that the λ-function, as defined by (5.20), is a modular function of weight 0 for $\Gamma(2)$.

(a) Verify from the definition in terms of the Dedekind η function that λ has weight 0.
(b) Find a finite set of matrices $G \subset \mathrm{SL}_2(\mathbf{Z})$ which are coset representatives for $\Gamma(2)$ inside $\mathrm{SL}_2(\mathbf{Z})$.
(c) Write each matrix in G in terms of the matrices $\left(\begin{smallmatrix} 1 & 1 \\ 0 & 1 \end{smallmatrix}\right)$ and $\left(\begin{smallmatrix} 0 & 1 \\ -1 & 0 \end{smallmatrix}\right)$.
(d) Prove that $\lambda(q)$ transforms correctly under each element of G; you may find the formula (5.5) useful.
(e) Why is λ not a modular function for $\Gamma_1(2)$? Why is it not a modular form?

(6) In this exercise we will find a formula for $r_8(n)$, the number of representations of the non-negative integer n as a sum of 8 squares, in terms of the coefficients of the Eisenstein series of level 4.

(a) Use Theorem 5.29 to show that $F(q) := \sum_{n=0}^{\infty} r_8(n) \cdot q^n$ is the Fourier expansion of an element of $M_4(\Gamma_0(4))$.
(b) Use the dimension formulae to show that $S_4(\Gamma_0(4)) = 0$, and to show that $\dim M_4(\Gamma_0(4)) = 3$.
(c) Find a basis for $M_4(\Gamma_0(4))$; you may wish to consider the various maps from $M_4(\mathrm{SL}_2(\mathbf{Z}))$ into $M_4(\Gamma_0(4))$. Write down the Fourier expansions of the elements of this basis explicitly.
(d) Write $F(q)$ in terms of your basis for $M_4(\Gamma_0(4))$ and equate Fourier coefficients to obtain a formula for $r_8(n)$.

(7) Let k be an integer. One standard application of the fact that $M_k(\mathrm{SL}_2(\mathbf{Z}))$ is a finite-dimensional vector space is to prove the identities (5.25), (5.26) and (5.27). We will now do this.

(a) Use the dimension formulae to show that the spaces of level 1 modular forms of weights 8, 10 and 14 are 1-dimensional.
(b) Find identities relating modular forms of weights 4, 6, 8, 10 and 14, and equate coefficients to prove the identities.

(8) In this question we will consider half-integral weight modular forms. Let k be a positive odd integer. Show that $M_{k/2}(\Gamma_0(4))$ has a basis of normalized eigenforms for all of the Hecke operators T_{p^2}.

(9) We will investigate classical modular functions of weight 0 and level N. To show that two modular functions are identical, we need to multiply both sides by a modular form with zeroes of the correct multiplicity to cancel the poles of the modular functions.

(a) Define $j_2(q) := j(q)/j(q^2)$. Show that this is a modular function of weight 0 for $\Gamma_0(2)$, and write j as a rational function in j_2.

(b) Define $j_4(q) := j(q)/j(q^4)$. Show that this is a modular function of weight 0 for $\Gamma_0(4)$, and write j_2 as a rational function in j_4.

(10) In this exercise we will prove Theorem 5.14 by computation. Choose a prime $p \in \{2, 3, 5, 7, 13\}$. (If you choose one of the larger p, you may wish to write a computer program to perform the computations).

(a) Check that the function $\Phi(q)$ is a modular function of weight 0 for $\Gamma_0(p)$.
(b) Show that the Fourier expansion of $\Phi(q)$ begins $q + \cdots$.
(c) Show that $j_p(q)$ is a modular function for $\Gamma_0(p)$.
(d) Successively subtract carefully chosen scalar multiples of $\Phi(q)^i$ from $j_p(q)$ until you have a modular function which has a zero of order at least p^2 at ∞.
(e) You should now have a function that vanishes to a very high order at ∞. Show that it is in fact identically zero by transforming the supposed identity of modular functions into an identity of modular forms.

(11) In equation (5.19), we have the "almost integer" $e^{\pi\sqrt{163}}$. We will now find some others.

(a) Find a list of the nine d such that the ring of integers of $\mathbf{Q}(\sqrt{d})$ has unique factorization.
(b) Use a reference such as [Silverman (1994)], Section II.6, and the formula (5.18) to show that the $1/q$ term is the dominant term.
(c) Use these to find more "almost integers".

(12) In this exercise we will show that $\tau(n) \equiv \sigma_{11}(n) \mod 691$.

(a) Compute the coefficients of the Eisenstein series E_{12}.
(b) Show that the space of modular forms that these modular forms live in is 2-dimensional.
(c) Find an equality relating $691E_{12}$, $691E_4^3$, and Δ.
(d) Reduce this equality modulo 691.

(13) Find similar congruences for other modular forms of level 1, using similar arguments (finding an equality of modular forms with integral coefficients, working out their coefficients, and reducing modulo N at an appropriate point).

(14) Show that if $k_1, k_2 \in \mathbf{N}$ and $k_1 \equiv k_2 \mod p^n(p-1)$ for some positive integer n, then $\sigma_{k_1}(m) \equiv \sigma_{k_2}(m) \mod$ for $m \in \mathbf{N}$.

(15) In this exercise we will check the details of the proof of Proposition 5.34.

(a) Verify that the formula given in (5.39) holds.

(b) Check that the six spaces $M_k(\Gamma_1(4))$, for $0 \leq k \leq 5$, are generated by monomials of the form $F^i(\theta^2)^j$ for $2i + j = k$.

(c) Show that $G_6 := F\theta^4(\theta^4 - 16F) \in S_6(\Gamma_1(4))$:

i. Firstly, check that G_6 is a modular form of weight 6.

ii. Secondly, prove that $h := \eta(q^2)^{12}$ is a cusp form for $\Gamma_1(4)$ of weight 6, by using the transformation formula (5.4), and by using the fact that $h^2 = \Delta(q^2)$ is a cusp form of weight 12.

iii. Finally, show that $h = G_6$ using the Sturm bound.

(d) Show that $G_5 := \theta^2(\theta^4 - 16F) \in S_5(\Gamma_1(4))$ by noting that as G_6 has zeroes at each cusp, then G_5 must have zeroes at each cusp also.

(e) Finally, use induction on the weight k to show that every modular form for $\Gamma_1(4)$ can be written as a polynomial in F and θ^2.

(16) In this exercise we will carry out the calculations to find the Fourier expansion of the unique normalized eigenform $h \in S_{10}(\Gamma_0(2))$, and then to represent it in terms of $F(q)$ and $\theta(q)^2$.

(a) Using the dimension formulae, or otherwise, show that $\dim S_{10}(\Gamma_0(4)) = 3$.

(b) Find a basis for the oldforms in this space in terms of h.

(c) Find a basis for the whole space $S_{10}(\Gamma_0(4))$ in terms of θ^2 and F.

(d) Combine these to find a representation of $h(q)$ in terms of $F(q)$ and $\theta(q)^2$, and check that your Fourier expansion for h matches that given in (5.40).

(17) Check the details from the section on CM forms. In particular:

(a) Check that the η-products given are actually modular forms.

(b) Identify the fields F which the modular forms f and g have complex multiplication by.

(c) Find an elliptic curve $E/\mathbf{Q}$ of conductor 32 and show that it has complex multiplication.

(18) Show that $\eta(q)$ is lacunary by estimating how many solutions to the equation $(3i^2 + i)/2 = n$ there are for integers in the range $1, \ldots, n$.

(19) Similarly, show that $\theta(q) = \sum_{n \in \mathbf{Z}} q^{n^2}$ is lacunary.

(20) In this exercise we will show that there are infinitely many CM modular forms for $\Gamma_1(4)$, with Fourier expansions of the form

$$f(q) = \sum_{m,n=-\infty}^{\infty} (m + in)^{k-1} q^{m^2+n^2}.$$

(a) Choose a suitable spherical function F with $\nu \geq 1$, using the Fourier expansion of the modular form f as a guide. (There is a hint for this given in the Hints and Solutions section of the book).
(b) Let k be a positive integer which is congruent to 1 modulo 4. Construct a theta function using F.
(c) Use Theorem 5.29 to show that the theta function you have constructed is a cuspform of weight k for $\Gamma_1(4)$. Verify that it is a CM form.
(d) What would happen if $k \equiv 3$ modulo 4?

(21) Substitute $x = q^3$ and $z = -y$ into the formula (5.46), and then replace q^2 by q to derive Euler's pentagonal number theorem from the Jacobi triple product identity.
(22) Use the techniques of Exercise 20 to create similar CM forms for $\Gamma_1(8)$.
(23) Show that $\eta(q^6)^4$ is a modular form in $S_2(\Gamma_0(36))$ and that it is lacunary.
(24) Finish the proof of Theorem 5.42 by showing that the function which maps the partition $\lambda_1 + \cdots + \lambda_r = n - b_i$ to either

$$(r + 3i - 1) + (\lambda_1 - 1) + \cdots + (\lambda_r - 1) = n - b_{i-1}, \text{ if } r + 3i \geq \lambda_1,$$

or

$$(\lambda_2 + 1) + \cdots + (\lambda_r + 1) + 1 + \cdots + 1 = n - b_{i+1}, \text{ otherwise,}$$

(where there are $\lambda_1 - r - 3i - 1$ 1s in the sum) is an involution, and hence a bijection, so (5.45) holds.

Chapter 6

Modular forms in characteristic p

In this chapter we will give a brief overview of the classical and modern theory of mod p modular forms, and indicate how one can compute with them. This can allow us to prove results about modular forms in characteristic 0, and is an interesting subject of study in its own right as well.

We will also give a brief overview of a very important conjecture relating Galois representations and modular forms; this is Serre's conjecture, introduced in [Serre (1987)]; this has been proved very recently, and we will discuss it in Section 6.2.

The classical treatment of mod p modular forms in [Serre (1973b)] and [Swinnerton-Dyer (1973)] views mod p modular forms as reductions of modular forms, in a way that we will explain in the next section. There are certain cases where this is not sufficient, and we will then briefly discuss a new type of modular forms which solve the problem; these are the Katz modular forms, which were introduced in [Katz (1973)].

Another standard reference is [Lang (1976)]; mod p modular forms are considered in Chapter X. The survey article [Ribet and Stein (1999)] gives an overview and some useful references to the literature.

6.1 Classical treatment

Firstly, we will prove some elementary results on the structure of the ring of mod p modular forms, and contrast this with the results that we have proved for characteristic 0 modular forms.

6.1.1 *The structure of the ring of mod p forms*

Let us consider modular forms for $\mathrm{SL}_2(\mathbf{Z})$ for simplicity; in this section, we will follow the exposition given in [Lang (1976)], Chapter X. We note that one can formulate similar results for congruence subgroups.

Throughout this chapter we will refer to the expansion in terms of q of a mod p modular form as the Fourier expansion of that modular form, to conform with our usage elsewhere.

Firstly, we will need to introduce some extra terminology. Let R be a commutative ring. We define $M_k(\Gamma; R)$ and $S_k(\Gamma; R)$ to be the spaces of modular forms (respectively, cusp forms) for the congruence subgroup Γ of weight k whose Fourier expansions are defined over R. Up to this point, we have always assumed that $R = \mathbf{C}$.

We recall from Proposition 3.6 that we can generate the algebra of modular forms for $\mathrm{SL}_2(\mathbf{Z})$ using the Eisenstein series E_4 and E_6; if we define $\mathcal{M}_R$ to be the graded ring of modular forms for $\mathrm{SL}_2(\mathbf{Z})$ whose Fourier expansions at ∞ have coefficients in R, we have that

$$\mathcal{M}_{\mathbf{C}} = \bigoplus_{k=0}^{\infty} M_k(\mathrm{SL}_2(\mathbf{Z}); \mathbf{C}) = \mathbf{C}[E_4, E_6],$$

and the graded ring of cusp forms for $\mathrm{SL}_2(\mathbf{Z})$ with Fourier coefficients in $\mathbf{C}$ (which is a graded ideal in $\mathcal{M}_{\mathbf{C}}$) is given by

$$\mathcal{S}_{\mathbf{C}} = \bigoplus_{k=0}^{\infty} S_k(\mathrm{SL}_2(\mathbf{Z}); \mathbf{C}) = \Delta\mathbf{C}[E_4, E_6].$$

Every modular form $f \in \mathcal{M}_{\mathbf{C}}$ can be written uniquely as a polynomial of the form

$$\sum_{4a+6b=k} \alpha_{a,b} \cdot E_4^a \cdot E_6^b \text{ with } \alpha_{a,b} \in \mathbf{C},$$

as E_4 and E_6 are linearly independent over $\mathbf{C}$ (see Exercise 5). A similar statement holds for cusp forms; we simply multiply through by Δ to enforce the requirement that the form vanishes at ∞.

If we wish to reduce the Fourier expansions of modular forms f defined over $\mathbf{C}$ modulo a prime p, then we will have to impose the conditions that f has a Fourier expansion defined over the rational numbers, and that the denominators of these coefficients are prime to p. Let $\mathbf{Z}_{(p)}$ be the local ring of $\mathbf{Z}$ at p; this is the set of rational numbers which have denominators prime to p. We will also need the ring $\mathbf{Z}[\frac{1}{2}, \frac{1}{3}]$; this is the ring of rational numbers with denominators of the form $2^a \cdot 3^b$. We see that these are reasonable conditions, at least for modular forms for $\mathrm{SL}_2(\mathbf{Z})$, because we have already proved that there is a $\mathbf{C}$-basis for $M_k(\mathrm{SL}_2(\mathbf{Z}))$, given by suitable powers of E_4 and E_6, whose elements have Fourier expansions defined over $\mathbf{Z}$. This means that we can find a basis of Hecke eigenforms which have Fourier expansions defined over the integers of some number field. We will see below that this basis does not give every modular form with integral Fourier coefficients.

For ease of notation in this section, we will drop the $\mathrm{SL}_2(\mathbf{Z})$ and refer to spaces of modular forms as M_k or $M_k(R)$ if we wish to make the ring of definition explicit.

We can consider the space of modular forms of weight k which have coefficients in $\mathbf{Z}_{(p)}$; these form a $\mathbf{Z}_{(p)}$-module $M_k(\mathbf{Z}_{(p)}) \subset M_k$. We will define $\overline{M}_k$ to be the *classical space of mod p modular forms (for* $\mathrm{SL}_2(\mathbf{Z})$*) of weight k*; it is the reduction of $M_k(\mathbf{Z}_{(p)})$ modulo p.

Over $\mathbf{C}$, we could generate the algebra of modular forms with E_4 and E_6, but we see that there are modular forms with integral coefficients which are not integral polynomials in E_4 and E_6, such as $\Delta = 1/1728(E_4^3 - E_6^2)$. There are two possible solutions to this: we can either add another generator, or consider modular forms over $\mathbf{Z}[\frac{1}{2}, \frac{1}{3}]$. Firstly, we will add a generator; we have already seen in Exercise 1 of Chapter 3 that we can find an integral basis using E_4, E_6 and Δ, and now we will find that basis explicitly. Let us recall Theorem X.4.3 from Lang:

Theorem 6.1. *Let k be an even integer and let $M_k(\mathbf{Z})$ be the space of modular forms of weight k with Fourier coefficients in $\mathbf{Z}$. Then $M_k(\mathbf{Z})$ has a $\mathbf{Z}$-basis, which is also a $\mathbf{C}$-basis for $M_k(\mathbf{C})$, of the form*

(1) $E_4^a \cdot \Delta^b$ with $4a + 12b = k$, if $k \equiv 0 \mod 4$,
(2) $E_6 \cdot E_4^a \cdot \Delta^b$ with $4a + 12b = k - 6$, if $k \equiv 2 \mod 4$.

Proof. There is a standard strategy for proof-by-induction of results about modular forms for $\mathrm{SL}_2(\mathbf{Z})$; we first show the desired result for some base case, normally for weights less than or equal to 12, and then we use multiplication-by-Δ to go to higher weights. We will follow this here.

For weight $k < 12$, we see that the dimension is 0 or 1, and therefore the result is obvious. The basis is either empty, or a product of the Eisenstein series E_4 and E_6, or the constant 1 (if $k = 0$).

So we now assume that $k \geq 12$, and that $f \in M_k(\mathbf{Z})$. There are two cases here; either f is a cusp form or it isn't. If f is a cusp form, then we have that f is the product of Δ and a modular form of weight $k - 12$, so $f \in \Delta M_{k-12}(\mathbf{C})$, and as Δ^{-1} has integral Fourier expansion coefficients, we see that $f/\Delta \in M_{k-12}(\mathbf{Z})$. We now use the induction hypothesis to write f/Δ as a polynomial with integral coefficients in E_4, E_6 and Δ, and we are done.

If f is *not* a cusp form, then we consider $f - a_0 \cdot E_4^a \cdot \Delta^b$ if $k \equiv 0 \mod 4$, or $f - a_0 \cdot E_6 \cdot E_4^a \cdot \Delta^b$ if $k \equiv 2 \mod 4$, for suitable choices of a and b. These are cusp forms of weight k, and so by dividing by Δ we can reduce our problem to the case of modular forms of weight $k - 12$. Therefore we are done. □

We note that there is another choice for a $\mathbf{Z}$-basis of $M_k(\mathrm{SL}_2(\mathbf{Z}))$, known as the "Victor Miller basis" (because it appears in Miller's thesis); it is also the reduced row echelon form of the basis. The proof given follows Theorem 4.4 of Chapter X of [Lang (1976)].

Proposition 6.2. *Let k be a positive integer. The space $M_k(\mathrm{SL}_2(\mathbf{Z}))$ has a basis $g_0, \ldots, g_r$, such that if $a_i(g_j)$ is the ith Fourier coefficient of g_j, then $a_i(g_j) = \delta_{ij}$. Also, the g_j have Fourier expansions defined over $\mathbf{Z}$.*

Proof. If we choose positive integers a, b such that $4a + 6b \leq 14$ and $4a + 6b \equiv k \mod 12$, then we can write

$$h_j = \Delta^j E_6^{2(r-j)+b} E_4^a;$$

we can check that this modular form has weight k, integral Fourier coefficients, and satisfies $a_j(h_j) = 1$ and $a_i(h_j) = 0$ if $i \leq j$. We can then use Gaussian elimination to find the g_j. □

We note that this basis is a good choice for computational purposes, because given a modular form in the space spanned by the Victor Miller basis, we can easily write it in terms of the basis elements.

The other choice is to change the basis ring; we consider modular forms with coefficients in $\mathbf{Z}[\frac{1}{2}, \frac{1}{3}]$; we invert these primes because they are the prime divisors of 1728, which is the denominator of Δ when it is written as a polynomial in E_4 and E_6.

Theorem 6.3. *Let k be an even integer and let $M_k(\mathbf{Z}[\frac{1}{2}, \frac{1}{3}])$ be the space of modular forms of weight k with Fourier coefficients in $\mathbf{Z}[\frac{1}{2}, \frac{1}{3}]$. Then we can write every $f \in M_k(\mathbf{Z}[\frac{1}{2}, \frac{1}{3}])$ as a polynomial in E_4 and E_6 with coefficients in $\mathbf{Z}[\frac{1}{2}, \frac{1}{3}]$.*

Proof. As before, we first check the base case; this time we need to consider weight 12 also. The spaces of modular forms with weights less than 12 are all one-dimensional, generated by an Eisenstein series or a constant. We can therefore write these as polynomials in E_4 and E_6.

For weight 12 we recall that $\Delta = 1/1728 \cdot (E_4^3 - E_6^2)$; this is where we need to invert 2 and 3.

For $k > 12$, we have either that f is a cusp form, in which case we consider f/Δ and use the induction hypothesis, or we consider $f - a_0 \cdot E_4^a \cdot E_6^b$ for a suitable choice of a and b; this will be a cusp form, so as before we can divide by Δ and reduce our problem to that of forms of weight $k - 12$. Thus we have proved the theorem as this is the induction hypothesis. □

We will now consider the structure of $M_k(\mathbf{F}_p)$. We have seen that $M_k(\mathbf{Z})$ can be generated by 3 generators (Δ, E_4, E_6), and that $M_k(\mathbf{Z}[\frac{1}{2}, \frac{1}{3}])$ can be generated by 2 generators (E_4, E_6). We will now prove the analogous result for $M_k(\mathbf{F}_p)$. There are two very different cases; $p < 5$ and $p \geq 5$; these are very different because the Fourier coefficients a_n for $n > 0$ of E_4 and E_6 vanish modulo 2 and modulo 3.

Theorem 6.4. *Let $p = 2$ or 3 and let k be an even integer. Then we have that*

$$\overline{E}_4 \equiv \overline{E}_6 \equiv 1 \mod p,$$

and

$$\mathcal{M}_{\mathbf{F}_p} = \mathbf{F}_p\left[\overline{\Delta}\right].$$

Furthermore, we see that $M_k(\mathbf{F}_p)$ is the set of polynomials in $\overline{\Delta}$ over $\mathbf{F}_p$ of degree at most r, where r is $\lfloor k/12 \rfloor$ if $k \mod 12 \neq 2$ and $\lfloor (k-2)/12 \rfloor$ if $k \mod 12 \neq 2$.

We now assume that $p \geq 5$. Then we have that

$$\mathcal{M}_{\overline{\mathbf{F}}_p} = \overline{\mathbf{F}}_p\left[\overline{E}_4, \overline{E}_6\right],$$

and every element of this ring can be written as a product of E_4, E_6 and $E_4^3 - \alpha E_6^2$, where $\alpha \in \overline{\mathbf{F}}$.

The space $M_k(\mathbf{F}_p)$ is given by the ring of graded polynomials in E_4 and E_6 of weight k.

We notice that if $p < 5$, then we can generate $M_k(\mathbf{F}_p)$ with *one* generator, Δ.

Proof. If $p < 5$, we see that

$$E_4(q) \equiv E_6(q) \equiv 1 \mod 2^4 \cdot 3,$$

so when we reduce modulo p we are left with the constants and Δ. We then use the dimension formula for characteristic 0 modular forms to prove the last assertion; recall that the dimension goes up linearly in the weight k, and the constant is $1/12$.

We now assume that $p \geq 5$. We can write an element of $M_k(\overline{\mathbf{F}}_p)$ as a graded polynomial in E_4 and E_6 (as we can invert 1728 in $\mathbf{F}_p$). Following the argument given in [Lang (1976)], we can set $E_4 = X^4$ and $E_6 = Y^6$, so a graded polynomial in E_4 and E_6 can be written as a graded polynomial F in X and Y.

We assume that there is a linear factor $X - \lambda Y$ of F over $\overline{\mathbf{F}}_p[X,Y]$. We see that if α is a 4th root of unity and β is a 3rd root of unity, then there is an automorphism of $\overline{\mathbf{F}}_p[X,Y]$ given by $X \mapsto \alpha X$ and $Y \mapsto \beta Y$ which fixes F. This means that

$$\prod_{\alpha,\beta}(\alpha X - \lambda\beta Y) = X^{12} - \alpha Y^{12} = E_4^3 - \alpha E_6^2$$

is a factor of F.

The assertion about $M_k(\mathbf{F}_p)$ follows by considering the structure of characteristic 0 modular forms of weight k over $\mathbf{Z}$ and then reducing them modulo p; we notice that we can write $\Delta = (E_4^3 - E_6^2)/1728$, and because we have assumed that $p \geq 5$ we can invert 2 and 3, so we can reduce $1/1728$ modulo p. □

One standard result about a particular modular form concerns the Eisenstein series E_{p-1}, when $p \geq 5$. We will prove the following result, which will allow us to determine the Fourier expansion of E_{p-1}. This will shed more light on the Bernoulli numbers, which we introduced in Section 2.3.

Proposition 6.5 (Von Staudt-Clausen congruence). *Let n be a positive integer and let $p \geq 5$ be a prime. Then*

$$pB_{p-1} \equiv 1 \mod p.$$

Proof. We note that this can be generalized to a result for all B_n, using the Kummer congruences (see Theorem X.1.1 of [Lang (1976)]), but we will not need that level of generality here. (The proof we give here was credited to Zagier by Lang).

If we consider the defining relation for the Bernoulli numbers and multiply both sides by a suitable multiplier, we have

$$\sum_{m=0}^{\infty} B_m \cdot \frac{t^m}{m!} = \frac{t}{e^t - 1}$$

$$\left(\frac{e^{pt}-1}{t}\right) \sum_{m=0}^{\infty} B_m \cdot \frac{t^m}{m!} = \frac{e^{pt}-1}{t} \frac{t}{e^t-1}$$

$$= \sum_{n=0}^{p-1} e^{nt}.$$

We now consider the coefficients of $t^m/m!$ on both sides of this equation. We find that

$$\sum_{m=0}^{p-1} \frac{(p-1)!}{(p-k)!k!} B_k \cdot p^{p-k} = \sum_{r=1}^{p-1} r^n,$$

and we extract pB_{p-1} from the left-hand side to obtain

$$pB_{p-1} = -\sum_{k=0}^{p-3} \binom{n}{k} pB_k \cdot \frac{p^{p-1-k}}{p-k} + \sum_{r=1}^{p-1} r^n.$$

Finally, we see using the fact that $p^{p-1-k}/(n-k+1)$ vanishes modulo p that pB_{p-1} has non-negative p-valuation, and hence that

$$pB_{p-1} \equiv \sum_{r=1}^{p-1} r^{p-1} \mod p;$$

using Fermat's Little Theorem we see that this sum is congruent to -1 modulo p, as required. □

This proposition will allow us to prove the following corollary about the Fourier expansion of E_{p-1}.

Corollary 6.6. *Let $p \geq 5$ be a prime number. Then*

$$E_{p-1}(q) \equiv 1 \mod p.$$

Proof. We apply Proposition 6.5 to the Fourier expansion of E_{p-1}, which we recall has the coefficient $2(p-1)/B_{p-1}$ in it. By the Proposition, we see that $1/B_{p-1}$ vanishes modulo p, so as $\sigma_{p-2}(n)$ takes values in $\mathbf{Z}$, we see that $E_{p-1}(q) \equiv 1 \mod p$. □

For $p = 2$ and $p = 3$, we can replace E_{p-1} by the Eisenstein series E_4, which vanishes modulo 2 and modulo 3. We notice that in this case we still have $p-1$ dividing 4.

We can in fact go further than this; for every prime p, there is a mod p modular form called the *Hasse invariant* which has Fourier expansion which is just the constant 1 and weight $p-1$.

For $p=2$ and $p=3$, this is *not* the reduction of a classical modular form with characteristic 0, because there are no classical modular forms for $\mathrm{SL}_2(\mathbf{Z})$ of weight or weight 2. The definition of the Hasse invariant in these cases is more technical; we can view the Hasse invariant as the reduction of a modular form for a congruence subgroup $\Gamma_1(N)$, where N is prime to p. We will find suitable Eisenstein series in Exercise 7.

Finally, we will state one case of what are known as the *Kummer congruences* relating different Bernoulli numbers. We will need this in the next section.

Proposition 6.7 (Kummer congruence). *Let p be a prime number, and let m and n be positive integers. If $m \equiv n \not\equiv 0 \mod (p-1)$, then*

$$\frac{B_m}{m} \equiv \frac{B_n}{n} \mod p.$$

Proof. There is a proof of this given as [Lang (1976)], Chapter X, Theorem 7.1, and another is given in [Ireland and Rosen (1990)], Chapter 15, Section 2, Theorem 5. □

6.1.2 *The θ operator on mod p modular forms*

We have seen in Chapter 2 that the θ operator "destroys modularity" (see Proposition 2.28); in other words, if f is a non-constant modular form, then $\theta(f)$ is *not* a modular form. However, in the mod p world we can define a θ operator which does preserve (mod p) modularity.

We extend the definition of the θ operator to be a formal derivative on power series over $\mathbf{F}_p$ in the natural way.

We now recall Theorem 7.1 from [Lang (1976)], which deals with the reduction of E_{p+1} modulo p.

Proposition 6.8. *Let $p \geq 5$ be a prime. Then*

$$E_{p+1}(q) \equiv E_2 \mod p, \text{ and} \tag{6.1}$$

$$\partial_k E_{p-1} \equiv E_{p+1} \mod p. \tag{6.2}$$

Proof. We note that $\sigma_p(n) \equiv \sigma_1(n)$ modulo p for every prime p, and that by Proposition 6.7, we have the correct congruence for the Bernoulli numbers, so (6.1) holds. We have therefore shown that θ preserves mod p modularity.

We also note that from the definition of ∂_k in (2.23) we have

$$\partial_k(E_{p-1}) = 12\theta(E_{p-1}) - (p-1)E_2E_{p-1}; \tag{6.3}$$

if we reduce both sides of this modulo p, and use that $E_{p-1} \equiv 1$ modulo p, and (6.1), then we have (6.2). □

By rearranging the defining equation for ∂_k, we see that, if $p \geq 5$, then

$$\theta(f) = \frac{\partial_k(f) + (k-1)E_2 \cdot f}{12};$$

we note that $E_2 \equiv E_{p+1}$ modulo p, and $E_{p-1} \equiv 1$ modulo p, so therefore $\theta(f) \equiv 1/12(\partial_k(f)E_{p-1} + (k-1)E_{p+1}f)$ is a mod p modular form of weight $k + p + 1$.

We note that if $p = 2$ or $p = 3$, then the space of mod p modular forms is given by the space $\mathbf{F}_p[\Delta]$, and that

$$\theta(\Delta) \equiv \Delta E_2 \mod p$$

because $\partial_{12}\Delta \in S_{14}(\mathrm{SL}_2(\mathbf{Z}))$, and this is a zero-dimensional vector space. We also note that $E_2 \equiv 1$ modulo 2 or 3, so therefore θ acts trivially on Fourier expansions modulo 2 or 3. This means also that it preserves modularity modulo 2 and 3, so we have showed that θ preserves mod p modularity for every prime p. As an illustration of this, let us consider the Δ function as a mod 13 modular form. We see that $g := \theta(\Delta)$ is a mod 13 modular form of weight 26 with no constant term, and by dimension considerations, we see that g must be the reduction of the unique cuspidal eigenform for $\mathrm{SL}_2(\mathbf{Z})$ of weight 26 (see Exercise 6). We note especially that this is a relation involving only (mod p) eigenforms.

6.1.3 *Hecke operators and Hecke eigenforms*

Hecke operators on mod p modular forms can be defined by taking the definition of Hecke operators on classical modular forms with coefficients in $\mathbf{Z}_{(p)}$ and reducing modulo p.

One interesting elementary observation is that the operators T_p and U_p that we have defined coincide modulo p if the weight is greater than 1. This is because the second factor in the definition vanishes; it contains a factor of p^{k-1} which is zero modulo p. On the other hand, if the weight is exactly 1 then the second term does *not* vanish. We will see in Section 6.3 that weight 1 is special in other ways.

We have seen that there are many congruences modulo p between modular forms of different weights, because there are modular forms such as E_{p-1} which are congruent to 1 modulo p. In [Jochnowitz (1982)], it is shown that there are only finitely many distinct systems of eigenvalues for any given congruence subgroup Γ. We note that this means that the Hecke operators acting on mod p modular forms are nonsemisimple.

In [Conrey *et al.* (2000)] it is shown that if we fix a congruence subgroup Γ then the characteristic polynomials of the Hecke operators acting on $S_k(\Gamma)$ are of the form

$$\prod_{i=1}^{n} f_i(x),$$

where the f_i are periodic.

Both of these results tell us that the Hecke operators and the Hecke algebra are rather different in the characteristic 0 and characteristic p worlds; for instance, it is conjectured that the characteristic polynomials of the Hecke operators acting on $S_k(\mathrm{SL}_2(\mathbf{Z}))$ are always irreducible (Maeda's conjecture; see Section 7.5.5).

6.2 Galois representations attached to mod p modular forms

We saw in the previous section that the Fourier expansion of a mod p modular form does not identify it uniquely; for instance, because the Eisenstein series E_4 and E_6 both vanish modulo 2, the mod 2 modular forms Δ and $\Delta E_4^a E_6^b$ have the same Fourier expansion.

This motivates us to define the *filtration* of a modular form f; the smallest weight at which there is a modular form with the same Fourier expansion as f.

Definition 6.9. Let p be a prime number and let $f \in M_k(\Gamma_1(N); \mathbf{F}_p)$. We define the *filtration* $w(f)$ to be the smallest non-negative integer $k' \leq k$ such that there exists $g \in M_{k'}(\Gamma_1(N); \mathbf{F}_p)$ such that $f(q) \equiv g(q) \mod p$.

We note here that the mod p filtration of the Eisenstein series E_{p-1} is 0, because it is congruent to the constant 1 which is a modular form of weight 0. In our motivating example above, we see that the mod 2 filtration of $\Delta E_4^a E_6^b$ is 12.

We see that if a mod p modular form f has weight k then its Fourier expansion appears as the Fourier expansion of a mod p modular form at weights $k+p-1$, $k+2(p-1)$, and so on, via multiplication by the Eisenstein series E_{p-1}. The converse to this is more difficult to prove.

One major motivation for the study of the arithmetic aspects of the Fourier expansions of modular forms is given by the following two theorems on Galois representations attached to modular forms. Firstly, we will briefly recall the definition of a *representation*; for our purposes, it is a continuous group homomorphism $\rho : G \to \mathrm{GL}(V)$, where G is a group and V is a vector space. A representation is said to be irreducible if it has no nonzero proper subrepresentations.

It is a standard result (see Chapter VI, Sections 1 and 2 of [Serre (1979)]) that there is an invariant N or N_ρ called the *Artin conductor* or simply the conductor of a representation ρ.

Let $K/\mathbf{Q}$ be a Galois extension with Galois group G. We recall that for a prime p which is unramified in K a *Frobenius element* Frob_p of a finite Galois group G is defined to be an element such that $\mathrm{Frob}_p(x) \equiv x^p \bmod p$; this is unique up to conjugacy class.

Theorem 6.10 (Deligne-Serre, [Deligne and Serre (1974)]). *Let $f \in S_1(\Gamma_1(N))$ be a normalized modular eigenform with Fourier expansion $f(q) = \sum_{n=1}^{\infty} a_n q^n$, with Dirichlet character ε such that $\varepsilon(-1) = -1$. Then there exists a finite Galois extension K of $\mathbf{Q}$ with Galois group G and a representation*

$$\rho_f : G \to \mathrm{GL}_2(\mathbf{C}),$$

such that, if $p \nmid N$, the characteristic polynomial of $\rho_f(\mathrm{Frob}_p)$ is

$$x^2 - a_p \cdot x + \varepsilon(p).$$

We note that the representation here is complex; in the following theorem it will be defined over $\overline{\mathbf{F}}_p$.

The following theorem was proved for weight 2 by Shimura, and for arbitrary weights greater than 2 by Deligne.

Theorem 6.11 (Shimura [Shimura (1966)], Deligne [Deligne (1973)]). *Let $f \in S_k(\Gamma_1(N); \overline{\mathbf{F}}_p)$ be a normalized modular eigenform of character χ with Fourier expansion $f(q) = \sum_{n=1}^{\infty} a_n q^n$. Let F be the field generated over $\mathbf{F}_p$ by the a_n. Then there exists a unique semisimple representation*

$$\rho_f : \mathrm{Gal}(\overline{\mathbf{Q}}/\mathbf{Q}) \to \mathrm{GL}_2(F)$$

such that

$$\mathrm{Trace}\, \rho_f(\mathrm{Frob}_l) = a_l$$

and

$$\det \rho_f(\mathrm{Frob}_l) = l^{k-1}\chi(l)$$

for $l \nmid pN$. *We say that* ρ_f is associated with f.

These results say that, given a modular form modulo p, we have a Galois representation. It is natural to ask if we have a result that goes the other way: given a Galois representation ρ, what conditions do we have to impose to be certain that there is a modular form f associated to ρ?

There is a famous conjecture of Serre which gives an answer to this:

Conjecture 6.12 (Serre, [Serre (1987)]). *Let p be a prime number and let* $\rho : \mathrm{Gal}(\overline{\mathbf{Q}}/\mathbf{Q}) \to \mathrm{GL}_2(\overline{\mathbf{F}}_p)$ *be an irreducible Galois representation. We say that ρ is* odd *if* $\rho(c) = -1$, *where c is the action of complex conjugation on* $\overline{\mathbf{Q}}$.

We say that a representation is modular (of weight k, level N and character χ) *if there exists a modular eigenform* $f \in S_k(\Gamma_0(N), \chi; \overline{\mathbf{F}}_p)$ *such that* $\rho_f \cong \rho$.

Assume that ρ is odd. Then we have the following conjectures:

(1) Weak Conjecture*:* ρ *is modular.*
(2) Strong Conjecture*:* ρ *is modular, of a weight, level and character that are given by an explicit recipe.*

Remark 6.13. We note that the Weak Conjecture is difficult to falsify; there is no obvious way for us to show that a given representation is not modular of *some* weight and level, although we can show that it is not modular of any given level. However, the Strong Conjecture is falsifiable; it gives a specific space of modular forms which we can compute and check to see whether there is a modular form with a suitable representation attached.

Remark 6.14. We note that it is necessary to consider mod p modular forms in this conjecture, because there exist cases where the space of modular forms specified in the recipe is trivial in characteristic 0 but nontrivial in characteristic p.

A good introduction to the theory of Serre's Conjecture as it was known before 2007 is the survey article [Ribet and Stein (1999)]. This explains things in a very down-to-earth way, with many explicit calculations.

Before 2007, the following was known:

Theorem 6.15. *Let F be a finite field of characteristic greater than 2 and let $\rho : \mathrm{Gal}(\overline{\mathbf{Q}}/\mathbf{Q}) \to \mathrm{GL}_2(F)$ be an irreducible representation. If ρ is modular, then ρ is modular of the given weight, level and character.*

This was also known for many cases when the characteristic was 2; one thing that makes the even characteristic case harder is that $-1 = 1$ in characteristic 2, so the condition of being an odd representation becomes trivial.

This means that the Weak and the Strong conjectures were known to be equivalent in many cases. The following was also known; in Section 6.7 of [Hellegouarch (2002)], it was said that "except for a finite number of special cases, it has not been proven in general":

Theorem 6.16. *Let ρ be an odd irreducible representation taking values in $\mathrm{GL}_2(\mathbf{F}_r)$ which satisfies some technical conditions. Then if $r \in \{2, 3, 4, 5, 7, 9\}$, ρ is modular.*

It should be noted that the theorem was known to be true unconditionally if $r = 2$ or $r = 3$, and that this fact plays a crucial role in the proof of Fermat's Last Theorem. It is also useful to note that these results were proved by methods which, although deep and ingenious, did not seem to generalize to larger finite fields; they were rather ad hoc. There was even speculation that Serre's conjecture could be false, as much of the numerical evidence known was for small finite fields, and could be explained without Serre's conjecture.

Recently, however, everything changed (see the introduction to Chapter 6 of [Hellegouarch (2002)], which talks about a "new paradigm" in the context of Fermat's Last Theorem; the new results on Serre's Conjecture were as startling, at least to this author). The following theorem was announced by Khare and Wintenberger and Kisin:

Theorem 6.17 (Serre's Conjecture for any conductor). *Let q be a prime power and let $\rho : G_{\mathbf{Q}} \to \mathrm{GL}_2(\mathbf{F}_q)$ be an irreducible odd representation. Then ρ is modular.*

In 2005, this theorem was proved for representations with Artin conductor 1 (that is, unramified away from p); see [Khare (2006b)] for the proof, and [Khare (2006a)] for a survey of the proof. The MathSciNet review calls this a "spectacular advance" over previous results, which it undoubtedly is, as it uses different methods which can be generalized to arbitrary finite fields. The theory of modularity is crucial to this; generalizations of the work of Wiles and Taylor in proving that representations are modular are an important ingredient of the proof of Serre's conjecture.

The references for the odd conductor case are [Khare and Wintenberger (2006a)] and [Khare and Wintenberger (2006b)]. The paper [Kisin (2008)] deals with the case of even conductor.

6.3 Katz modular forms

In Section 6.1.1, we defined mod p modular forms to be the reductions of classical modular forms. However, we recall that we remarked that Serre's conjecture in certain circumstances predicts the existence of a mod p modular form f of weight 1, but where there is no characteristic 0 modular form of weight 1 that can be reduced mod p to give f. We insist on weight 1 for this point, because if we allow higher weights then we can lift f to characteristic 0.

Work of Mestre (published as Appendix B in [Edixhoven (2006)], but carried out during the 1980s), generalized by Wiese (published as Appendix A in [Edixhoven (2006)]), shows that there are examples where this actually occurs; when $p = 2$, there are circumstances when conjectures predict that there should be a mod 2 modular form of weight 1, but there is no suitable weight 1 form in characteristic 0, but there is a suitable mod 2 object. Mestre's original example was a mod 2 modular form f of level 1429 and weight 1 with Fourier expansion defined over $\mathbf{F}_8$ which begins

$$f(q) = q + \alpha q^2 + \alpha q^3 + (\alpha^2 + 1)q^4 + (\alpha^2 + \alpha)q^5 + O(q^6), \tag{6.4}$$

where $\alpha^3 + \alpha^2 + 1 = 0$. This cannot be the reduction of a characteristic 0 modular form of weight 1 because the image of the Galois representation ρ associated to f is $\mathrm{SL}_2(\mathbf{F}_8)$, and there is no finite subgroup G of $\mathrm{GL}_2(\mathbf{C})$ such that G has $\mathrm{SL}_2(\mathbf{F}_8)$ as a quotient.

The solution to this problem is to expand the definition of mod p modular form to include forms which do not come from characteristic 0. This requires us to define modular forms as rules which map pairs consisting of an elliptic curve E defined over a ring R and a "level N structure", a map $\mathbf{Z}/N\mathbf{Z} \to E[N]$ satisfying some compatibility properties. This follows ideas introduced by Katz in [Katz (1973)]. It can be shown that the space of modular forms given by this definition when $R = \mathbf{C}$ is isomorphic to the definitions we gave in Chapter 2. The notation $S_k(\Gamma; R)_{\mathrm{Katz}}$ is often used for spaces of Katz modular forms to make it clear that one is referring to these and not to classical modular forms.

The advantage of considering modular forms as rules acting on objects in a certain way which satisfy certain compatibility properties is that we can generalize the fields that they are defined over. Also, they provide a more conceptual reason for viewing the study of modular forms as an area of number theory rather than of complex analysis; it is an interesting feature of the way modular forms theory is normally taught that first one considers certain holomorphic functions on $\mathcal{H}$, and only after this does one encounter Hecke operators and their arithmetic. Using the approach of Katz we can define modular forms over a general commutative ring R without having to ensure that R is a subring of $\mathbf{C}$, and one can begin with a number-theoretic ring without having to begin with complex functions.

The result of Mestre and others that we mention above is a good example of how computational power can be used to find examples in number theory; to find mod 2 weight 1 modular forms which cannot be lifted to characteristic 0, one performs a computer search and then checks for modular forms whose associated Galois representation has large enough image H so that there is no finite subgroup G of $\mathrm{GL}_2(\mathbf{C})$ such that G has H as a quotient. Having a fast computer means that one can find large numbers of examples.

We will briefly sketch one way to find mod p weight 1 modular forms that is used in [Edixhoven (2006)]. We note that the Hasse invariant has weight $p-1$, and that raising to the pth power in characteristic p comes down to mapping q to q^p. We therefore look for a power series $f(q)$ such that both $f(q)$ and $f(q^p)$ are the Fourier expansions of mod p modular forms of weight p. This can be efficiently done by a computer; we will see an easy example of this in Section 6.5. This also relies on us being able to say definitively that a Fourier expansion is of the form $g(q^p)$; we quote the following theorem:

Theorem 6.18 ([Edixhoven (2006)], Proposition 4.2). *Let p be a prime, let χ be a Dirichlet character modulo N, and define*

$$B := \frac{p+2}{12} N \prod_{\substack{l \mid N \\ l \text{ prime}}} \left(1 + l^{-1}\right).$$

If $g \in S_p(\Gamma_1(N), \chi; \overline{\mathbf{F}}_p)$ has Fourier expansion given by $g(q) = \sum_{n=1}^{\infty} a_n q^n$ with $a_n = 0$ for all n less than B which are not divisible by p, then $g = f(q^p)$ and there exists a unique modular form $f \in S_1(\Gamma_1(N), \chi; \overline{\mathbf{F}}_p)$ with Fourier expansion $f(q)$.

We note that the $p+2$ in the definition of B appears because the proof uses the properties of the θ operator divided by the Hasse invariant, which raises the weight by $p+1-(p-1)=2$.

6.4 The Sturm bound in characteristic p

In Chapter 4, we saw (using Proposition 4.24) that we could find a basis whose elements are eigenforms for the Hecke operators T_p (for $p \nmid N$) for the space of modular forms $S_k(\Gamma_1(N))$. We can also prove that these bases have Fourier expansions which are defined over the integers of a certain number field. This means that we can reduce the Fourier expansions of characteristic 0 modular forms modulo a suitable prime ideal.

We have the following version of the Sturm bound "mod p", which allows us to check whether the reductions of two modular forms of the same weight and level are identical using a finite amount of computation. This is in fact the original version of the Sturm bound given in [Sturm (1987)]. A modern reference for this is [Stein (2007)], Section 9.4.

Theorem 6.19. *Let K be a number field and let $\mathcal{O}_K$ be its ring of integers. Let $\mathfrak{m}$ be a prime ideal in $\mathcal{O}_K$, and let Γ be a congruence subgroup of* $\mathrm{SL}_2(\mathbf{Z})$ *of index M. Suppose that $f \in M_k(\Gamma)$ is a modular form whose Fourier expansion $f(q)=\sum_{n=0}^{\infty} a_n q^n$ takes values in $\mathcal{O}_K$, and that $a_i \in \mathfrak{m}$ for $i \in \{0,1,\ldots,kM/12\}$. Then $a_n \in \mathfrak{m}$ for* all *$n \in \mathbf{N}$.*

Remark 6.20. We see that this theorem implies the characteristic 0 version of the Sturm bound, because we can apply Theorem 6.19 for every maximal ideal $\mathfrak{m}$ to obtain the characteristic 0 Sturm bound (recall that an integer that is divisible by infinitely many distinct prime numbers must be 0).

Proof. The proof strategy is very similar to that used for the Sturm bound in Chapter 3. We first show that it is true for $\mathrm{SL}_2(\mathbf{Z})$ and then reduce the general case to this special case. Here, we will prove it for $\mathrm{SL}_2(\mathbf{Z})$ and leave the general case as an exercise.

We will use the Δ function, which as we recall has integral Fourier expansion beginning $q-24q^2+O(q^3)$. We recall that it has a unique zero at the cusp ∞ and no other zeroes.

The order of vanishing of $G := f^{12}/\Delta^k$ at ∞ is at least $-k$, because f is holomorphic at ∞. We also see that the order of vanishing of G modulo $\mathfrak{m}$ is strictly greater than 0, because the order of vanishing of Δ^{-k} modulo $\mathfrak{m}$ is at most k, and the assumption in the theorem is that the order of vanishing of f modulo $\mathfrak{m}$ is strictly greater than $k/12$.

These two observations allow us to write its Fourier expansion as

$$\frac{f^{12}}{\Delta^k} = \sum_{n=-k}^{\infty} a_n q^n,$$

where the a_n are all in the ring of integers of the field K, and if $n \geq 0$ then we have that $a_n \in \mathfrak{m}$.

Recall that Proposition 5.5 says that we can write the modular function G as a polynomial in the j-invariant of degree at most k. This means that we have

$$\frac{f^{12}}{\Delta^k} \in \mathfrak{m}[j],$$

with degree at most k, and because Δ^k has integral coefficients, we see that $f^{12} \in \Delta^k \cdot \mathfrak{m}[j]$.

Therefore, we have that f^{12} vanishes modulo $\mathfrak{m}$, and because $\mathfrak{m}$ is a prime ideal, we see that this means that f vanishes modulo $\mathfrak{m}$, as required, so every Fourier coefficient of f lies in $\mathfrak{m}$.

The proof for general congruence subgroups Γ follows the idea given in the characteristic zero proof; we find the cosets

$$\mathrm{SL}_2(\mathbf{Z}) = \bigcup_{i=1}^{m} \Gamma\gamma_i,$$

(where without loss of generality we may take $\gamma_1 = 1$) and consider the modular form

$$F := f \cdot \prod_{i=2}^{m} f|[\gamma_i]_k.$$

One can show that F is a modular form for $\mathrm{SL}_2(\mathbf{Z})$, and then apply the result we have proved above. □

6.5 Computations with mod p modular forms

We notice that the area of mod p modular forms is an especially suitable topic for computation; if we wish to check whether two modular forms have Fourier expansions which are congruent modulo a prime number p, then we can find the Sturm bound N, compute both Fourier expansions to the precision q^N and then check if their difference really does vanish modulo p. Let us consider an explicit example.

It can be shown by consulting Cremona's tables of modular elliptic curves [Cremona (1997)] that there are three isogeny classes of elliptic curves of conductor 503. The associated cuspidal modular eigenforms in $S_2(\Gamma_0(503))$ have Fourier expansions beginning

$$f_1(q) = q + q^2 + q^3 - q^4 - 2q^5 + q^6 + O(q^7)$$
$$f_2(q) = q + q^2 + 3q^3 - q^4 - 2q^5 + 3q^6 + O(q^7)$$
$$f_3(q) = q - q^2 + q^3 - q^4 - 4q^5 - q^6 + O(q^7).$$

We can check that for this space of modular forms the Sturm bound is 84, and by explicit computation we can see that the Fourier expansions of $f_1 - f_2$, $f_1 - f_3$ and $f_2 - f_3$ vanish modulo 2 up to this bound, so the three modular forms are pairwise congruent modulo 2. In Exercise 2 we will check this statement explicitly. This example is taken from [Kilford (2002)].

We now consider another example where we relate modular eigenforms of different weights and levels. Let us consider the Δ function and the modular eigenform $f \in S_6(\Gamma_0(7))$ which has Fourier expansion

$$f(q) = q - 10q^2 - 14q^3 + 68q^4 - 56q^5 + O(q^6).$$

We will show that

$$f(q) \equiv \Delta(q) \mod 7. \tag{6.5}$$

We first multiply f by the Eisenstein series E_6 to obtain a modular form of the same weight as Δ. By Corollary 6.6, we recall that $E_6(q) \equiv 1$ modulo 7. We then consider the modular form $g := \Delta - f \cdot E_6$, which is an element of $S_{12}(\Gamma_0(7))$.

The Sturm bound for this particular space of modular forms is 8, so we need only check that the first 8 Fourier coefficients are congruent to 0 modulo 7 to apply Theorem 6.19 to g. We invite the reader to verify this in Exercise 3.

We can discover many more congruences of this general form; for instance, we can show that the Fourier expansion of the (unique) normalized eigenform $f \in S_2(\Gamma_0(11))$ is congruent to $\Delta(q)$ modulo 11, by the same method of multiplying by a suitable Eisenstein series; see Exercise 4.

Finally, we present an example of how to find a mod p weight 1 modular form. Let $N = 23$ and $p = 2$. The space $S_2(\Gamma_0(23))$ can be shown to be 2-dimensional, with basis

$$f(q) = q - q^3 - q^4 - 2q^6 + 2 + q^7 - q^8 + O(q^9)$$
$$g(q) = q^2 - 2q^3 - q^4 + 2q^5 + q^6 + 2q^7 - 2q^8 + O(q^9).$$

We see that (to $O(q^9)$) we have that $f(q)^2 \equiv g(q)$ modulo 2. By Theorem 6.18, we see that this is enough accuracy to be able to show that $f(q^2) \equiv g(q)$ modulo 2, as $B = 8$ in this case, and therefore that there is a modular form $h \in S_2(\Gamma_1(23); \mathbf{F}_2)$ with Fourier expansion equal to $f(q)$.

Similar computations can be carried out for other levels and primes p; in Exercise 8 we will verify Mestre's computation of a mod 2 weight 1 form which is not the reduction of a characteristic 0 weight 1 form.

6.6 Exercises

(1) Prove Theorem 6.19 for an arbitrary congruence subgroup. There are proofs of this given in Sturm's original paper [Sturm (1987)] and in [Stein (2007)], Chapter 9.

(2) Check the assertions given in the example concerning $S_2(\Gamma_0(503))$; compute the Sturm bound and check that the forms f_1, f_2 and f_3 are congruent modulo 2 to a sufficiently high degree of accuracy.

(3) Check that the congruence (6.5) holds, using a computer algebra package of your choice, or otherwise.

(4) In this exercise we will show that the modular forms $\Delta \in S_{12}(\mathrm{SL}_2(\mathbf{Z}))$ and $f = \eta(q)^2\eta(q^{11})^2 \in S_2(\Gamma_0(11))$ are congruent modulo 11.

   (a) Find an Eisenstein series E which is congruent to 1 modulo 11 and has weight 10.

   (b) Compute the modular form $\Delta - f \cdot E$ to a suitable precision to show that its Fourier expansion at ∞ vanishes modulo 11.

(5) Prove that the Eisenstein series E_4 and E_6 are linearly independent over $\mathbf{C}$.

(6) Check the assertion given in Section 6.1.2 that

$$\theta(\Delta) \equiv \Delta E_4^2 E_6 \bmod 13$$

and find other congruences of the form $\theta(f) \equiv g$ such that both f and g are eigenforms.

(7) Let $p \in \{2, 3\}$. We will find suitable modular forms which are characteristic 0 lifts of the Hasse invariant.

   (a) Let $p = 2$. Show that the theta function with Fourier expansion

$$E(q) := \sum_{x,y \in \mathbf{Z}} q^{2(x^2+xy+y^2)} = 1 + 6q^2 + 6q^6 + O(q^7)$$

   is a modular form for $\Gamma_1(12)$ of weight 1, and that its reduction modulo 2 has Fourier expansion 1.

(b) Now let $p = 3$. Show that the modular form F given by

$$F(q) := 2E_2(q^2) - E_2(q)$$

is a modular form for $\Gamma_0(2)$ of weight 2, and that its reduction modulo 3 has Fourier expansion 1.

(8) Prove using the methods outlined in Sections 6.3 and 6.5 that there is a mod 2 weight 1 modular form of level 1429 with Fourier expansion given by (6.4).

Chapter 7

Computing with modular forms

I wish to God these calculations had been performed by steam. — Charles Babbage, 1812, on finding many errors in tables compiled by the Royal Astronomical Society.

A modern reference for those who want to calculate with modular forms is Stein's textbook [Stein (2007)]. This is unusual in that it emphasizes the computational aspects of the subject, rather than the theory.

Other references for computations in number theory are [Crandall and Pomerance (2005)], the books [Cohen (1993)] and [Cohen (2000)] on algorithms and computational number theory. The comprehensive survey given by [Dickson (1966)] also includes some information about historical computations.

7.1 Historical introduction to computations in number theory

There has been a long history of computation in pure mathematics generally and number theory specifically, and in the area of modular forms. In the Exercises following Chapter 1 of [Hellegouarch (2002)], a passage from Jakob (I) Bernoulli's *Ars Conjectandi* of 1713 is quoted, where he uses properties of the Bernoulli numbers, which we defined in Chapter 2, to compute the sum of the tenth power of the first 1000 positive integers; he says *"[t]his example shows the uselessness of the book* 'Arithmetica infinitorum' *by Ismael Bullialdus, which is entirely devoted to a tremendously large computation ... [which is] less than what I accomplished here in a single page"*. To quote the entry for 107,928,278,317 in [Wells (1998)], mathematicians "*can be as competitive as athletes.*"[1]

Possibly the oldest computations which can be seen to be in the field of modular forms are the classical Greek computations of the *figurate numbers* (Pythagoras is credited with initiating the study of these numbers); these are numbers which can be represented as a pattern of dots; for instance, there are the triangular numbers $1, 3, 6, 10, \ldots, n(n+1)/2, \ldots$, the square numbers $1, 4, 9, 16, \ldots, n^2, \ldots$, the pentagonal numbers $1, 5, 12, 22, \ldots, n(3n-1)/2, \ldots$, and so on for polygons of more and more sides. Figure 5.2 shows a graphical representation of the first few pentagonal numbers. One of the many great triumphs of Euler was to show that the following relationship holds (we proved this as Theorem 5.43):

$$\prod_{n=1}^{\infty}(1-q^n) = \sum_{m=-\infty}^{\infty}(-1)^m q^{m(3m-1)/2}.$$

We note that the left hand side of this equality is the Dedekind η function divided by $q^{1/24}$, which is where the link to the theory of modular forms comes from. The right hand side can be seen to give the generalized pentagonal numbers (these are the values of $n(3n-1)/2$ for $n \in \mathbf{Z}$).

Another computation from classical antiquity which can be fitted into the framework of modular forms is one performed by Diophantus of Alexandria, who worked during the 3rd century AD. He has been called "the father of algebra", and gives his name to the solution of polynomial equations in integers, which are known as *Diophantine equations*, although he often found rational solutions to the equations he considered. In book VI of the *Arithmetica* ([Sesiano (1982)]), he finds x such that $x^3 - 3x^2 + 3x + 1$ is a square; this is essentially finding rational points on a modular elliptic curve associated to a modular form in $S_2(\Gamma_0(1728))$ (it can be shown that there are infinitely many points defined over $\mathbf{Q}$, generated by the point $(0, 1)$).

We have mentioned that modular functions can be used to compute digits of π. In the 17th century, Isaac Newton wrote in a letter that "I am ashamed to tell you to how many figures I carried these computations [of the digits of π], having no other business at the time".

In the 18th century, Gauss was well-known for his calculating ability, as well as his more theoretical work, such as the justly famous *Disquisitiones arithmeticae* [Gauss (1986)]. His rediscovery of the asteroid Ceres in 1801

[1] This number is interesting because it is the first in an eighteen-term arithmetic progression of primes, with common difference 9922782870. It has fairly recently been proved that there are arbitrarily long arithmetic progressions consisting only of primes; see [Green and Tao (2008)].

was aided by his ability to calculate logarithms in his head without using pencil and paper. He encouraged the prodigy Zacharias Dase to perform number-theoretic calculations, such as the computation of digits of π (using formulae similar to (5.22) to calculate π *in his head*) and extending tables of integer factorizations; Dase computed the factorizations of the integers between 7000000 and 10000000.

Babbage, in the 19th century, designed mechanical computers (the Analytic Engine and the Difference Engine) which would have been able to compute many interesting number-theoretic quantities, such as the Bernoulli numbers, if they had ever been completed. He did complete smaller mechanical devices, which were capable of enumerating the values of polynomials such as $f(n) = n^2 + n + 41$, which has the property that $f(i)$ is prime if i is an integer between 0 and 40.

The work of Ramanujan in the early 20th century is still important today; there is a journal named after him which publishes papers in the areas that he worked on[2]. One of his great strengths was in calculation, which gave him evidence upon which he could make his conjectures, which were backed up by a very strong intuition for what should be true: for instance, the Ramanujan conjecture on the size of the coefficients of the Δ function. We have seen in Theorem 5.43 a proof of this result.

J. W. L. Glaisher [Glaisher (1907a,b,c)] performed calculations at the beginning of the 20th century for certain modular forms related to the representation of positive integers as the sum of certain numbers of squares, and gave some tables of the mathematical functions he calculated. It is interesting to note how much more complicated the formulae become as the number of squares increase. It is interesting to note the lengths of these papers, which take up most of one volume of the journal; the first few cases (representations of an integer as the sum of 2,4 or 6 squares) take only a few lines or at most a page, while the final case, which deals with representations of an integer as the sum of 18 squares, requires over 60 pages. Both Ramanujan (in [Ramanujan (1916)]) and Louis Mordell (in [Mordell (1918)]) subsequently extended this work.

During the latter part of the 19th century, through to the middle of the 20th century, there was a great interest in the creation of tables by mechanical means. There is a project which is still running, called the Cunningham Project, which aims to factor certain integers of the special type $b^n \pm 1$. The latest versions of these tables can be found in [Brillhart

[2]This is a very rare honour.

et al. (2002)] and at their website[3]. One body which did much to promote the creation of tables was the British Association Mathematical Tables Committee; Glaisher, who we mentioned above, was the first secretary of this body. Cunningham left a substantial amount of money to create number-theoretic tables; one of these which has some relation to modular forms is the table of representation of primes by quadratic forms [Gupta *et al.* (1960)], as these can be viewed as determining the coefficients of theta functions. There were other volumes which are more closely related to the theory of modular forms; one was a table of partitions, and another gave values of the Riemann ζ function. A full account of the history of this committee can be found in [Croarken and Campbell-Kelly (2000)], and [Campbell-Kelly *et al.* (2003)] is a general history of mathematical tables.

Although the following devices have very little to do with modular forms computations, they are striking enough to be worthy of a brief review. One of the most unusual primality-proving devices was that built by Lehmer in the 1930s; it used bicycle chains with gaps of different lengths to find integers which satisfied congruence conditions. In [Lehmer (1928)] he gives details of its operation, and how it could factor $N := (10^{20}+1)/(10^4+1)$ into the product of two primes. There were other innovative analogue computers built; the Phillips Economic Computer (see [Swade (1995)]) simulated the economy of a developed country by hydraulic means, by using coloured water pumped between tanks.

In the 1960s, Atkin was performing calculations at the Atlas Computer Laboratory; according to Birch's memoir of this time [Birch (1998)], one of the uses of modular form theory at the time was to compute $\eta(q)^{26}$ to test the performance of their computer hardware; this is an interesting application of modular forms, comparable to the use of computations of digits of π in the manner of Algorithm 5.16. Atkin's work at the time dealt with congruences between modular forms, in the style of Ramanujan, and also on the subject of modular forms for non-congruence subgroups.

In [Atkin (1968)] (quoted in [Birch (1998)]) the following sentence appears: "I can admit the possession of 2 cwt. of computed tables, of which at least 2 stone I have not even thought worth examining". The era of access to large-scale computer power has meant that it is now very easy to, in Swinnerton-Dyer's words (also quoted in [Birch (1998)]) "pile up lists of integers in the manner of a magpie" which are "neither designed to produce

---

[3] This can be found at `http://homes.cerias.purdue.edu/~ssw/cun/`.

valuable results nor capable of doing so".

On the other hand, the Birch–Swinnerton-Dyer conjecture, which relates the rank of an elliptic curve E over a number field to the order of vanishing of its L-function (an analytic object associated to E and analogous to the Riemann ζ function), was motivated by a large computation performed on the EDSAC computer at Cambridge, which had been used earlier to find the then largest prime number known ($180 \cdot (2^{127} - 1)^2 + 1$, in 1951). This was also supported later on by evidence for elliptic curves with special properties, for instance those defined over $\mathbf{Q}$ with complex multiplication.

There were a series of conferences in Antwerp during the 1970s [Kuyk (1973); Deligne and Kuyk (1973); Kuyk and Serre (1973); Birch and Kuyk (1975)] on modular forms in one variable. These included computational papers, and some tables of modular forms, which are sometimes called the "Antwerp tables" in the literature.

Computations with weight 2 modular forms were often motivated by the study of modular elliptic curves. One of the direct descendants of the tables of elliptic curves given in the proceedings of the Antwerp conferences are the tables of elliptic curves over the rational numbers by Cremona [Cremona (1997)]. A related table is reported on in [Stein and Watkins (2002)]; this contains about 44 million elliptic curves, including nearly all curves with conductor less than 10^8.

In the 1990s, a package called HECKE for doing computations with modular forms, via the arithmetic of modular symbols, was developed by William Stein. This allows us to calculate modular forms (at least with weight $k \geq 2$) in a systematic way, thus unifying and generalizing those tables which had been compiled beforehand.

We will now introduce some mathematical software which can be used to do computations with modular forms, including descendants of the package HECKE that we have just mentioned.

7.2 MAGMA

The MAGMA system [Bosma *et al.* (1997)] was developed at the University of Sydney and released in 1994; the University releases new versions and updates on a regular basis. Its predecessor, Cayley, was first described in 1981–82 (see [Cannon (1984)]). MAGMA is available under a non-commercial cost-recovery license.

The recent volume [Bosma and Cannon (2006)] gives a selection of examples of papers which use MAGMA to investigate mathematical problems. MAGMA has a large package, based on William Stein's package HECKE, supporting computations on modular forms; finding eigenforms, computing bases, computing Hecke operators. We will discuss some of these here.

MAGMA can be invoked from the command line in Linux or Unix using the command magma if it is installed locally.

```
ljpk@magma:~$ magma
Magma V2.14-1    Thu May 22 2008 12:09:00 on magma
[Seed = 854292760]
Type ? for help.  Type <Ctrl>-D to quit.
```

We now define a space of modular forms:

```
> S12:=CuspidalSubspace(ModularForms(1,12));
> S12;
Space of modular forms on Gamma_0(1) of weight 12 and
dimension 1 over Integer Ring.
> Basis(S12);
[
    q - 24*q^2 + 252*q^3 - 1472*q^4 + 4830*q^5 - 6048*q^6
      - 16744*q^7 + O(q^8)
]
```

This is the modular form $\Delta(z)$ that we defined previously. We will now compute 500 terms of the q-expansion of Δ, and print out the 499th term.

```
> Delta:=qExpansion(Basis(S12)[1],500);
> Coefficient(Delta,499);
-10887719272500
```

We can also compute Eisenstein series:

```
> M6:=ModularForms(1,6);
> EisensteinSeries(M6);
[*
    -1/504 + q + 33*q^2 + 244*q^3 + 1057*q^4 + 3126*q^5
          + 8052*q^6 + 16808*q^7 + O(q^8)
*]
```

and we can check that $\Delta = (E_4^3 - E_6^2)/1728$ by the following computation:

```
> M4:=ModularForms(1,4);
```

```
> E4:=qExpansion(EisensteinSeries(M4)[1],500);
> E6:=qExpansion(EisensteinSeries(M6)[1],500);
> ((240*E4)^3-(-504*E6)^2)/1728-Delta;
O(q^500)
```

By calculating the Sturm bound in this case using Theorem 3.13, we see that a modular form of level 1 and weight 12 which vanishes to order 500 must be identically zero. In fact, we can compute the Sturm bound (which we recall is also known as the Hecke bound):

```
HeckeBound(ModularForms(1,12));
1
```

This indicates that our computation to that precision was not actually necessary, but it is good to check that the program is producing the correct output.

We can also compute Hecke operators. Let us illustrate with the example of $S_2(\Gamma_0(37))$:

```
> S37w2:=CuspidalSubspace(ModularForms(37,2));
> HeckeOperator(S37w2,3);
[ 1  0]
[ 2 -3]
```

and we can find eigenforms for the Hecke operators too:

```
> Newforms(S37w2);
[* [*
    q - 2*q^2 - 3*q^3 + 2*q^4 - 2*q^5 + 6*q^6 - q^7 + O(q^8)
*], [*
    q + q^3 - 2*q^4 - q^7 + O(q^8)
*]*]
```

We notice that the eigenvalues of the matrix of the Hecke operator T_3 are the Fourier coefficients a_3 of these newforms.

We can also find and verify congruences between two modular forms. Here, we define the ring of power series over $\mathbf{F}_{691}$ and then show that Δ and E_{12} are congruent modulo 691 (using Theorem 6.19):

```
> PSR<q>:=PowerSeriesRing(FiniteField(691));
> E12:=qExpansion(EisensteinSeries(M12)[1],500);
> PSR!(E12-Delta);
O(q^500)
```

We note that we can go in the other direction; given a power series which is the Fourier expansion of a modular form to a given precision, we can create that modular form. (The notation `$1` used here means "the result of the last computation").

```
> PSR<q>:=PowerSeriesRing(Rationals());
> S37w2!(q-2*q^2+O(q^3));
q - 2*q^2 - 3*q^3 + 2*q^4 - 2*q^5 + 6*q^6 - q^7 + O(q^8)
> Parent($1);
Space of modular forms on Gamma_0(37) of weight 2 and dimension
over Integer Ring.
```

In fact, it can be shown that the fragment of Fourier expansion given suffices to identify this modular form uniquely, using the Sturm bound.

7.2.1 MAGMA *philosophy*

We notice from the discussions above that there are some unusual features in the MAGMA system; we had to define the power series rings over $\mathbf{F}_{691}$ and $\mathbf{Q}$ before we could use them. This occurs more generally; MAGMA has strong typing, so one needs to define objects before they are used. Polynomial rings also know what they are defined over, and the system will give an error if one uses elements of the wrong ring:

```
> PR<x>:=PolynomialRing(Rationals());
> 1.2*x;
>> 1.2*x;
      ^

Runtime error in '*': Bad argument types
Argument types given: FldReElt, RngUPolElt[FldRat]
> Parent(1.2);
Real field of precision 30
```

Here we see that we cannot automatically use real numbers, like 1.2, as coefficients of a polynomial over the rationals. If we wish to use this, we must consider it as a rational number:

```
> Parent(6/5);
Rational Field
> 6/5*x;
6/5*x
```

There is, however, some automatic coercion between different fields; for instance, one can coerce rational numbers into the reals as follows (because there is an injection $\mathbf{Q} \hookrightarrow \mathbf{R}$):

```
> 4321/8765+0.0;
0.492983456930975470621791215060
```

We can also coerce objects from one structure to another using the `!` operator:

```
> E:=EllipticCurve([0,1]);
> E![2,3];
(2 : 3 : 1)
```

We note that the name MAGMA is derived from the Bourbaki definition of a *magma* to be a set M equipped with a single binary operation $M \times M \to M$.

The error messages given by MAGMA can also be quite interesting; consider

```
> Universe({});
>> Universe({});;
Runtime error in 'Universe': Illegal null set
```

7.2.2 MAGMA *programming*

Although MAGMA has many functions already defined, it is quite common that one wishes to do something new. There are two ways that one can define new functions in MAGMA. If we are writing a short program, then we can use the `function` terminology:

```
> increment:=function(n);return n+1;end function;
> increment(3);
4
```

This is easy to implement and is suitable for simple functions, but if we are writing something that we intend to use more generally, then we can write functions using the `intrinsic` command; some examples are given as Appendix A. Intrinsics must be attached before they are used:

```
> Attach("AppendixA.m");
> ModularFormsDerivative(Newforms(S37w2)[1][1]);
10*q + 4*q^2 - 54*q^3 - 148*q^4 - 500*q^5 - 348*q^6
     - 706*q^7 + O(q^8)
```

```
> Parent($1);
Space of modular forms on Gamma_0(37) of weight 4 and dimension 1
over Integer Ring.
```

Once a file has been attached to MAGMA, it will be monitored by the system, and if it has been changed then its contents will be re-loaded into memory when a MAGMA command is run.

There are several advantages to using intrinsics when one is programming in MAGMA: they check the types of the input, one can "overload" them (which means that we can have several different possible input types, which is not possible with functions)[4], and their signature can give useful information on what they do:

```
> ModularFormsDerivative;
Intrinsic 'ModularFormsDerivative'
Signatures:
    (<ModFrmElt> f) -> ModFrmElt
        This computes the partial_k map from M_k to M_(k+2).
```

This is much more informative than the information MAGMA stores for a function:

```
> increment;
function(n) ... end function
```

This helps in maintaining good documentation, which is important if one wishes to reuse old code or share code with other people; in particular, the type-checking encourages good programming practice. One can set MAGMA up so that it attaches the user-defined intrinsics when one runs MAGMA.

On the other hand, if we have a plain text file with MAGMA commands in it, we can use the `load` command to load and run these. We can also output text to a file for future reference, as well as saving the whole MAGMA state to be reloaded at a later date.

We also note that MAGMA stores the results of computations; if we compute `ModularFormsDerivative` with the same inputs twice, the second time will be almost instantaneous, as it can recall the result of the first computation from memory. This particular feature can become problematic if one runs a large number of computations in a single session, as this will

[4]It is instructive to type `ModularForms;` at the command-line to see just how many different ways there are to construct such spaces.

use more and more memory; the standard way to get around this is to write a batch file to run each computation in its own MAGMA process.

7.3 SAGE

The SAGE [Stein and Joyner (July 2005)] computational algebra package is a new open-source project led by William Stein which supports computations with modular forms (as well as many other things). The modular forms commands are very similar to those of MAGMA; we now demonstrate:

```
ljpk@sage:~/% sage
----------------------------------------------------------------
| SAGE Version 3.0.1, Release Date: 2008-05-05                 |
| Type notebook() for the GUI, and license() for information.|
----------------------------------------------------------------
sage: S24=ModularForms(1,24).cuspidal_subspace()
sage: S24
 Cuspidal subspace of dimension 2 of Modular Forms space
 of dimension 3 for Congruence Subgroup Gamma_0(1) of weight 24
 over Rational Field
sage: S24.hecke_polynomial(3)
 x^2 - 339480*x - 19020146544
```

SAGE has an object-oriented syntax, like that in languages such as C++. This means that there are specific functions available for objects of a particular type; here the space of modular forms has about 200 available functions, amongst which are the computation of the Hecke polynomials, the cuspidal subspace, the dimension and many others. We can also compute the Hecke operators as matrices:

```
sage: S24.hecke_matrix(7)
[     -707479910536 -1396411809408/19]
[     6796980449280        706120726136]
```

and we can compute the Hecke algebras too:

```
sage: S24.hecke_algebra()
Full Hecke algebra acting on Cuspidal subspace of dimension 2
of Modular Forms space of dimension 3 for Congruence Subgroup
Gamma0(1) of weight 24 over Rational Field
```

We can even chain together lots of functions, although this may not be a sensible idea if we are trying to understand what we are doing:

```
sage: S24.hecke_polynomial(7).factor_mod(5)
 (x + 1) * (x + 4)
```

This computes the factorization over $\mathbf{F}_5$ of the characteristic polynomial of T_7 acting on $S_{24}(\Gamma_0(1))$.

We can also find a basis of modular forms for a space of cusp forms; we now exhibit the two eigenforms of level 37 and weight 2:

```
sage: S37=ModularForms(37,2).cuspidal_subspace()
sage: S37_basis=S37.basis()
sage: f=S37_basis[0]
sage: g=S37_basis[0]-2*S37_basis[1]
sage: f
q + q^3 - 2*q^4 + O(q^6)
sage: g
q - 2*q^2 - 3*q^3 + 2*q^4 - 2*q^5 + O(q^6)
```

This computes the two eigenforms to precision q^6. We can compute these to arbitrary precision. Note that the elements of the list are indexed from 0, unlike in MAGMA, which indexes from 1. There is a great deal of functionality in the Hecke algebra implementation; one can compute dimensions; one can compute the Hecke bound and the Hecke algebra:

```
sage: S37.sturm_bound()
 7
sage: S37.hecke_algebra()
 Full Hecke algebra acting on Cuspidal subspace of dimension 2
 of Modular Forms space of dimension 3 for Congruence Subgroup
 Gamma0(37) of weight 2 over Rational Field
```

SAGE also understands the standard congruence subgroups $\Gamma_0(N)$ and $\Gamma_1(N)$. It can, for instance, compute sets of coset representatives for these in $\mathrm{SL}_2(\mathbf{Z})$:

```
sage: G=Gamma0(3)
sage: list(G.coset_reps())
 [[1, 0, 0, 1], [0, -1, 1, 0], [0, -1, 1, 1], [0, -1, 1, 2]]
```

We can find out what commands are available for a given object by the following construction:

```
sage: G.
G.Hom    G.__reduce__     G._maple_    G.generators
[more output omitted]
```

This in fact produces a vast list of output; there are many functions available for a generic group.

7.3.1 SAGE *philosophy*

One of the core features and philosophies of SAGE is that one can call functions from other packages, such as PARI (installed by default) and MATHEMATICA (as long as it is installed locally). The project in its own words is about "building the car instead of reinventing the wheel"; to this end, they incorporate over 50 other open-source packages (such as GAP, PARI, and Singular). This can be very useful if one wishes to use packages or reuse one's own code written in those languages. Here is an example of this, which computes the size of the torsion subgroup of the elliptic curve $y^2 = x^3 + 1$, as well as a generator of it:

```
sage: gp('elltors(ellinit([0,0,0,0,1]))')
 [6, [6], [[2, 3]]]
```

We note also that SAGE, like MAGMA, has strong typing; we must define polynomial and power series rings before they are used (the terminology for this is very similar, although not identical, to that of MAGMA):

```
sage: (t^2-1).factor()
 NameError: name 't' is not defined
sage: PR.<t>=PolynomialRing(Rationals())
sage: (t^2-1).factor()
 (t - 1) * (t + 1)
```

It is also free and open-source; it is released under the General Public License and is intended to be extended by those who use it.

7.3.2 SAGE *programming*

SAGE is implemented in the computer language Python; this is described by the project as an advantage, because it is a well-known and well-documented language independently of its use in their package, so there are reference sources available to support it, such as [Lutz (2003)]; it is also available under a free license, so it also is open-source. Other computer algebra packages have defined their own custom languages that one programs in.

SAGE shares some of the syntax of MAGMA; in particular, there is the same distinction between `load` and `attach`, in that a text file which is loaded into SAGE is simply run, whereas the contents of an attached file are updated if the file is later updated.

There is a similar construction to the MAGMA `function` command in SAGE:

```
sage: def add(a,b):
 ....:        return a+b
sage: add(10,12)
 22
```

We note that because SAGE uses Python the indentation given here is important.

We can also define the equivalent of MAGMA intrinsics in SAGE. Because it is designed to interface with many other systems, we can write it in many ways: as a SAGE script, or in Python, GAP, PARI, or in C++. There is a version of Python used by SAGE called Cython which can be much faster, and was specifically designed with speed in mind. One can even run MAGMA or MATHEMATICA code in SAGE, if one has these programs installed locally.

7.3.3 *The* SAGE *interface*

One innovative feature of SAGE, which we will now discuss, is the notebook interface. MAGMA and PARI both have a command-line interface, and while SAGE also has a command-line interface, it also has a notebook mode (like that of MATHEMATICA or MAPLE). Unlike those systems, however, this is accessed using a web browser such as Firefox or Internet Explorer. This fits in with the SAGE philosophy that, where possible, functionality should be re-used and not coded from scratch; using already-designed and well-documented web browsers means that their design features can be employed, and designers will be more familiar with off-the-shelf web browsers than a custom interface.

The fact that web browsers are used here means that the graphical version of SAGE can be used remotely, which is an unusual and potentially very useful feature for educational purposes; it also allows users to use the graphical interface with a more powerful remote computer.

7.3.4 SAGE *graphics*

Although the systems we have discussed are command-line based, they can give graphical output. One application of this for SAGE is that it can be used for teaching calculus, where one would like to graph functions over the real numbers, for instance, to help students visualize the material they are learning.

Figure 5.1 was created using the `plot` commands built in to SAGE; we give the code that was used in the Hints and Answers section at the back of the book. The syntax for these commands, in the words of the SAGE team (in the Reference Manual), was "inspired by the interface for plotting in Mathematica"; one can define elements separately and then add them together to create an image:

```
sage: line1=line([[0,0],[10,10]])
sage: line2=line([[0,0],[-10,10]])
sage: (line1 + line2).save("twolines.png")
```

This produces the image shown in Figure 7.1. We can add colour to these plots by using `rgbcolor` with each part.

7.4 PARI and other systems

7.4.1 PARI

There are many other computer algebra packages available. PARI [Pari] (also known as GP or PARI/GP) is free and open-source (released under the General Public License); it was originally developed by Henri Cohen and his co-workers at the University of Bordeaux 1.

Although PARI does not have as many built-in functions for modular forms as MAGMA and SAGE, it can be used for some tasks. We recall that the Δ function is the 24th power of the Dedekind η function, and use PARI to calculate it (we have to add in the extra q because PARI has $\eta(q)$ defined so as to exclude the $q^{1/24}$):

```
ljpk@meccah:~/% gp
                     GP/PARI CALCULATOR Version 2.3.3 (released)
[snipped]
parisize = 4000000, primelimit = 500000
? \ps5
   seriesprecision = 5 significant terms
```

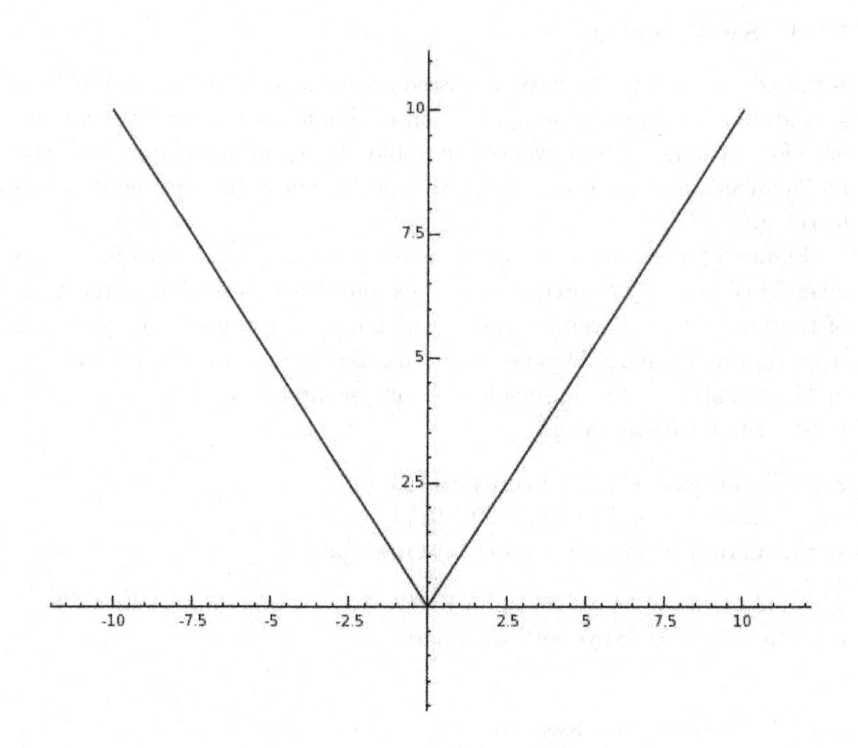

Fig. 7.1 A sample SAGE image

```
? q*eta(q)^24
%1 = q - 24*q^2 + 252*q^3 - 1472*q^4 + 4830*q^5 + O(q^6)
```

Programming in PARI is much like programming in C, so it may appeal to those with a background in this. It lacks strong typing, which is such a feature of MAGMA and SAGE; this may or may not appeal to the reader. The elliptic curves support is good, which may be very important for some modular forms calculations.

PARI can also produce visual output in the form of plots of functions; the syntax for the ASCII plot is `plot(X=a,b,expr,{Ymin},{Ymax})`, which produces a rendition of the graph of `expr` (this can even cope with trigonometric functions), and `ploth(X=a,b,expr)` gives a high-precision plotting of `expr`. One can create Postscript versions of these by using the command `psploth`.

In Figure 7.2, we see the graph of abs(x) between -10 and 10, as produced by PARI.

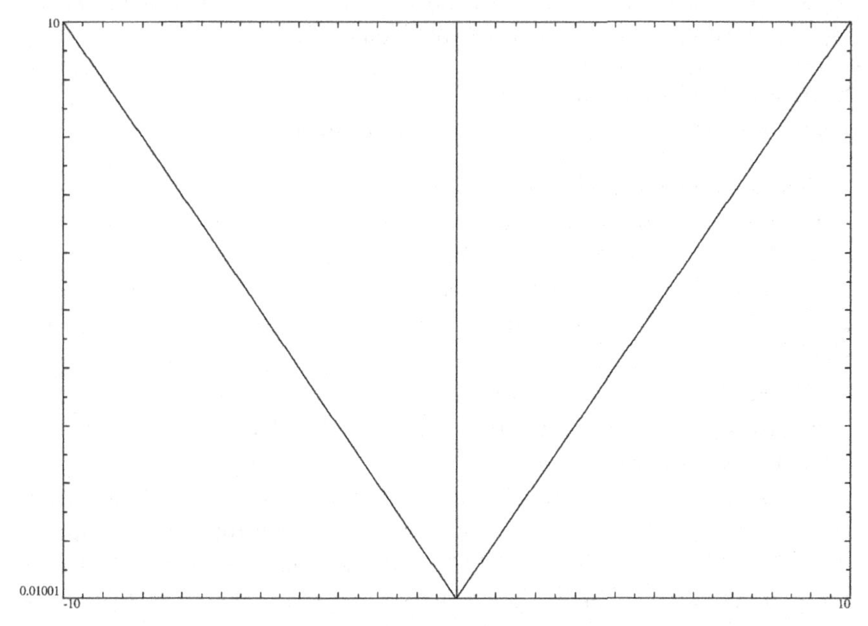

Fig. 7.2 A sample PARI image

There are books which use PARI as their computer algebra package of choice; for instance [Rodriguez Villegas (2007)], and others which recommend it, such as [Cohen (1993)], [Cohen (2000)] and [Rose (1994)]. It has also been used for modular forms computations; [Coleman *et al.* (1998)] includes a PARI program as its Appendix.

7.4.2 *Other systems and solutions*

There have been a few modular forms libraries written for the graphical computer algebra packages MAPLE[5] [Heck (1993)], MATHEMATICA[6] [Wolfram (1996)] and MATLAB[7] [Hunt *et al.* (2006)], such as Garvan's package QSERIES [Garvan (1998)] or Grotendorst's package [Grotendorst (1991)], but these programs are not optimized for number theoretic computations, and do not have the codebase of MAGMA or SAGE. MAPLE, MATHEMATICA and MATLAB are all non-free software, although many universities and

[5]Maple is a registered trademark of Waterloo Maple Inc.
[6]Mathematica is a registered trademark of Wolfram Research, Inc.
[7]MATLAB is a registered trademark of the MathWorks Inc.

companies have license arrangements for at least one of these.

Much of the source code of MAPLE (although not the kernel) can be viewed by the user using the following commands:

```
> interface (verboseproc=2);
> print (plot); # or another function name
```

These graphical packages can be very useful for more informal computations; for instance, the author's computations of modular functions of level 2^n in terms of the modular j-invariant were greatly assisted by MAPLE. Computations that do not require either a lot of memory or a lot of predefined functions may also be easier to implement in a graphical package.

Helena Verrill has an online package to draw pictures of fundamental domains, which is written in Java [Verrill (2001)]. This is also available as MAGMA code and in C.

There is a trade-off between using more standard packages and getting the very best performance. If one really wants to exploit the computer power available, then writing in C or in FORTRAN may be the best strategy; there are standard reference books like [Press *et al.* (2002)] to help write fast and efficient code. The downside of this approach is that it may be very difficult to make stand-alone packages work with each other; one may spend more time fiddling with the output of one program to make it acceptable as input to the next than actually solving the problem. Also, even a relatively unspecialized general package may be faster than a specially-written piece of code that has not been optimized.

7.5 Discussion of computation

7.5.1 *Computation today*

We live in a Golden Age of computational mathematics. Numerical problems that once seemed completely intractable are now trivial. A good reference for this section is [Crandall and Pomerance (2005)].

In [Shanks (1993)], page 195, the following quote is given, from Peter Barlow's *A New Mathematical and Philosophical Dictionary* (London, 1814):

> [$2^{30}(2^{31}-1)$] is the greatest perfect number known at present, and probably the greatest that ever will be discovered; for as they are merely curious, without being useful, it is not likely that any person will at-

tempt to find one beyond it.

This quote is often used in historical surveys of the literature; it is an example of a statement ("Clarke's First Law") made by the science fiction author Arthur C. Clarke in [Clarke (1962)], that

> [w]hen a distinguished but elderly scientist states that something is possible, he is almost certainly right. When he states that something is impossible, he is very probably wrong.

It is not clear what lessons we can learn from the history of such statements as Barlow's, except possibly that it is wise not to make unambiguous and unqualified statements about the future.

A good example of the progress made in computation over the 20th century is the factorization of the Mersenne number $2^{67} - 1$, which was conjectured to be prime by Mersenne. In 1903 it was shown to be the product of the primes 193707721 and 761838257287 by F. N. Cole [Cole (1903)]; he presented this as a (silent) talk at a meeting of the American Mathematical Society in 1903 (he calculated 2^{67} and subtracted 1 on the blackboard, then multiplied the two primes together and showed that these two calculations gave the same answer). This factorization, according to his account (as reported in the entry for $2^{67} - 1$ in [Wells (1998)]), took *3 years* of Sundays to find. By comparison, both MAGMA and SAGE on a modern desktop find the factorization in about a tenth of a second; see Exercise 3.

Another representative example is given in the answer to Exercise 1.2.4 of [Knuth (1969b)]; an account of the computation of 1000! by Uhler (published as [Uhler (1956)]); this was a project spanning several years and requiring much use of a hand-calculator, but it now takes less than a second on a modern laptop computer to compute.

There was also a period of 75 years, between 1876 and 1951, when the largest known prime number did not change (it was $2^{127} - 1$, which has 39 digits). The advent of the age of electronic computers changed this; during the last two decades, the largest known prime number changes approximately every other year. As of 2008, the largest known prime number is $2^{32582657} - 1$, which has just under 10 *million* digits. For more information on prime number records, the Prime Pages [Caldwell (2007)] are an informative and frequently updated resource.

Computers which are much more powerful than those which helped get the Apollo missions to the surface of the Moon and back are now routinely installed in washing machines and mobile phones. However, it is still quite important to do calculations in as efficient a manner as possible, in order to be able to push one's calculations as far as possible.

Also, many algorithms have an expected running time which is a non-linear polynomial in some parameter. Inefficiencies in such an algorithm can multiply and make it very difficult to run the computations in a reasonable time; one standard problem is that a certain command is in the centre of a set of nested loops and must therefore be run many times; if this is slow it will slow down the whole computation. We will now explain what this means.

7.5.2 *Expected running times*

The literature often says that an algorithm "runs in polynomial time". What does this mean? Here is a formal definition.

Definition 7.1. Let $f(x)$ be a polynomial in x. If n is the degree of f, we say that *f is $O(x^n)$* (it is "big O of x^n).

Let n be the size of the input to a problem. If the time taken for an algorithm A to run on this input is a polynomial in n (so is $O(n^k)$) then we say that A runs in *polynomial time*, or *A is in P*, where P is the class of algorithms that run in polynomial time.

A good example of an algorithm that has only been proved to run in polynomial time recently is primality testing; given an integer n, is n prime or not? The problem is normally formulated in terms of $\log(n)$; the number of digits of n. In 2002, it was finally proved that primality testing of integers was in P, by Agrawal, Kayal and Saxena [Agrawal *et al.* (2004)]. Their original preprint gave a deterministic algorithm that proved primality unconditionally in time $O((\log n)^{12} \cdot f(\log\log n))$, where f is a polynomial.

It is common that the first, pioneering, algorithm given to perform some computation does not have the best possible running time; this is certainly true for the AKS algorithm, although it was a superb piece of work and used comparatively elementary techniques. Indeed, the published version of their paper has running time $O((\log n)^{15/2} f(\log\log n))$ (a suggestion of Hendrik Lenstra helped improve the running time), and there are variants of their test (such as [Berrizbeitia (2005)]) which have proved running time $O((\log n)^6)$ which work for numbers of certain form.

It is important to realize that there are many useful algorithms which are not deterministic; for instance, the most effective primality tests known are either not deterministic, or do not have proved bounds. The fastest known currently is the elliptic curve primality proving method (ECPP) [Atkin and Morain (1993)], which has a heuristically estimated running time of $O((\log n)^{4+\varepsilon})$.

There is another class of algorithms which is of great interest; the *NP class*; these are problems for which an answer can be checked in polynomial time. It can be seen that P is a subset of *NP*.

One very important example of a problem known to be in *NP* but not known to be in P is the problem of factoring an integer N. If we have a claimed factorization of N as ab, then we can check it in polynomial time by multiplying a and b together. However, there is no known polynomial time algorithm for finding a nontrivial a and b. There is a \$1,000,000 prize on offer from the Clay Mathematics Foundation for a proof either that $P = NP$ or that $P \neq NP$; this is one of the Millennium Prize Problems (see [Devlin (2002)] for a gentle introduction to the problems). This particular problem is interesting even to non-mathematicians because the RSA algorithm for cryptography relies on the difficulty of factoring an integer $N = pq$, where p and q are (unknown!) primes.

7.5.3 *Using computation effectively*

In Cassels's book [Cassels (1991)], the following enlightening passage appears in Chapter 26:

> ...the problem of factorizing an integer n is constructive. All one has to do is to test all integers $m < n^{1/2}$ for divisibility. When, say, n has 100 decimal digits, this could take longer than the age of the universe.

If we are to perform calculations, we must carefully choose the method that we are using; two methods which are equivalent in a theoretical sense may have very different running times and memory requirements. For instance, we can write the Fourier expansion of the ubiquitous modular form Δ in the following two ways:

$$\begin{aligned}\Delta(q) = \eta(q)^{24} &= q\prod_{n=1}^{\infty}(1-q^n)^{24}\\ &= \frac{691}{762048}\left(E_{12} - E_6^2\right)\\ &= \frac{691}{762048}\left(\left(1 + \frac{65520}{691}\sum_{n=1}^{\infty}\sigma_3(n)q^n\right) - \left(1 + \sum_{n=1}^{\infty}\sigma_5(n)q^n\right)^2\right).\end{aligned}$$

However, because we have to raise one of these to a 24th power and the other only to a second power, the amount of computational time that we have to spend on each can be very different.

On the other hand, if we are seeking to perform a strictly limited computation, it may be more reasonable not to optimize our code, if the optimization will be time-consuming and difficult. This is not an exact science; every case must be considered on its own merits. There is a famous quote of Donald Knuth, creator of TEX: *"Premature optimization is the root of all evil"* ([Knuth]). Another example where practical considerations are important is in finding bases of eigenforms for spaces of modular forms. It is quite possible that we can find many choices of Hecke operators $\{T_p\}$ which we can use to find a basis of eigenforms (we decompose the space into eigenspaces for each T_p in turn, until the spaces are all 1-dimensional). However, we will normally use the Hecke operators T_2, T_3, $T_5, \ldots$ in order, as these are easiest to compute on Fourier expansions.

There are modern books which deal with effective algorithms which can be implemented on a computer; Cohen's books [Cohen (1993)] and [Cohen (2000)] provide details on algorithms which are useful for number theory in general, and [Stein (2007)] gives details of algorithms for modular forms.

7.5.4 *The limits of computation*

There are certain areas of the theory of modular forms where computations can give definitive answers; we can show that two modular forms are or are not congruent modulo p, or exhibit a basis of eigenforms for a specific space of modular forms, for instance. However, we cannot prove the general theorem that $S_k(\mathrm{SL}_2(\mathbf{Z}))$ has a basis of eigenforms purely by computation. This is a general theme in mathematics; we can disprove a conjecture by computation, but we cannot prove general theorems by computing examples.

One useful feature of spaces of modular forms of weight k that we have used throughout this book is that they are finite-dimensional, so that one can, for instance, find the basis of eigenforms with a finite amount of computation. This is not the case for the space of modular functions of weight k, which is infinite-dimensional. If a modular form vanishes to a sufficiently high order at ∞, then it must be identically zero, but this is not true for a modular function; just consider

$$j(z)^{-10^{100}} = q^{10^{100}} + \cdots ;$$

we cannot say that a given modular function of weight 0 is zero just because it has a zero of order 10^{100} at ∞. For comparison, it is estimated that there are about 10^{80} elementary particles in the Universe, so no naïve computation will be able to detect that this modular function is nonzero.

There are also examples of disproved conjectures which were based on large amounts of computational evidence, but which do not hold in general. For instance, there were conjectures of Gouvêa and Mazur on the roots of the characteristic polynomials of Hecke operators. To state them, we will need some notation.

Definition 7.2. Let $g(x) = \sum_{i=0}^{n} a_i x^i$ be a polynomial with coefficients in an extension K of $\mathbf{Q}_p$. Let v_p be the valuation on K, normalized such that $v_p(p) = 1$. The *p-slopes* of g are the slopes of the line segments of the lower convex hull of the points $\{(i, v_p(c_i))\}$.

If the prime p is clear from the context, we will talk about the *slopes of g* without qualification.

Let N be a positive integer. We define $d(k, \alpha)$ to be the number of roots of the Hecke polynomial T_k acting on modular forms of weight k and level Np which have slope α. For instance, if $N = 1$ and $p = 3$, then $d(12, 2) = 1$ and $d(12, \alpha) = 0$ for all other α, because the Fourier expansion of the unique modular form in $S_{12}(\mathrm{SL}_2(\mathbf{Z}))$ begins $q - 24q^2 + 252q^3 + O(q^4)$..

Conjecture 7.3 ([Gouvêa and Mazur (1992)], Conjecture 1). *Let α be a positive rational number and let k_1, k_2 be integers such that $k_1, k_2 \geq 2\alpha + 2$. If $k_1 \equiv k_2 \mod p^n(p-1)$ for some integer $n \geq \alpha$, then $d(k_1, \alpha) = d(k_2, \alpha)$.*

This conjecture was made based upon extensive numerical evidence; work of Mestre was summarized in section 5 of [Gouvêa and Mazur (1992)], and in the subsequent work of Wan [Wan (1998)] a weaker version of the conjecture with $k_1 \equiv k_2 \mod p^{m(n)}$, where $m(x)$ is a quadratic polynomial depending only on Np. Another system of conjectures was given by Buzzard in [Buzzard (2005)], and another conceptual system of conjectures was given by Clay in [Clay (2005)].

However, in [Buzzard and Calegari (2004)], a counterexample was found. Let $p = 59$ and $N = 1$. Then there exists a rational α such that $0 \leq \alpha \leq 1$ and $d(16, \alpha) \neq d(3438, \alpha)$; as $3438 = 16 + 58 \cdot 59$, this means that the Gouvêa-Mazur conjecture, as currently stated, is false. Buzzard and

Calegari also found other computational counterexamples.

Another famous number-theoretic example of a numerically testable conjecture that was eventually proved to be false is the Mertens Conjecture. Let μ be the Möbius μ function; the Mertens Conjecture is that

$$\left|\sum_{k=1}^{n} \mu(k)\right| < \sqrt{n}$$

for every positive integer n. For every n that has ever been tested explicitly, this statement is true, but it was proved in [Odlyzko and te Riele (1985)] that the conjecture is false, although the proof does not give an explicit n for which it fails. It is known that there is a counterexample with $10^{14} \leq n \leq \exp(3.21 \times 10^{64})$, but this range is far too large to search with current computers.

7.5.4.1 *Explicit examples of limitations*

We now consider some examples of how computer algebra systems can give non-intuitive answers.

For this example, we need to know that the p-adic ring $\mathbf{Z}_p$ is given by the direct limit

$$\mathbf{Z}_p := \lim_{\leftarrow} \mathbf{Z}/p^k\mathbf{Z}$$

and the MAGMA command `Zero` gives the zero element of a ring. A gentle introduction to the p-adic numbers is given by [Gouvêa (1997)]; we also note that $\mathbf{Z}_p$ is the completion with respect to (p) of the ring $\mathbf{Z}_{(p)}$ that we saw in Section 6.1.1. We now run a fairly natural computation:

```
> Z2:=pAdicRing(2);
> IsZero(Zero(Z2));
false
```

The issue here is that this particular representation of the 2-adic numbers is not exact; it does not actually have a zero element, just approximations to it. We see that a specific request for a zero element returns neither an error, nor an element which is a true zero.

A more familiar example is that of the real numbers given to a fixed precision (the floating-point numbers); it can happen that addition fails to be associative, in the following way. Let us assume that we are working with floating-point numbers defined to ten decimal digits of precision, and wish to compute the sum $1000000001 - 1000000003 + 2.511111111$. Then we find that

$$\begin{aligned}
&(1000000001 - 1000000003) + 2.511111111 \\
&= -2 + 2.511111111 = 0.511111111 \\
&\neq 1000000001 + (-1000000003 + 2.511111111) \\
&= 1000000001 - 1000000000 = 1.
\end{aligned}$$

This follows because there are inaccuracies caused by the loss of precision when we add the very large and very small numbers together; here this loss of precision is catastrophic, as our two answers agree to only one decimal digit of precision.

For more details of exactly how floating-point numbers are represented within a computer, we refer to Section 4.2.2 of [Knuth (1969a)]; the example we have just given is based on the one given by Knuth.

7.5.5 *Guy's law of small numbers*

Computations also have the problem that they are biased towards "small" numbers, as the dimension of spaces of modular forms goes up with the weight and the level. In [Guy (1988)], Guy introduces the notion of the "Law of Small Numbers"; there are not enough small numbers for all of the tasks that they are set to do, so that spurious patterns emerge when one considers insufficiently many members of a set (for instance, a set of spaces of modular forms).

It may be very hard to gather data on the behaviour of (say) Hecke algebras of high dimension, as the computations get harder and harder as the level and weight increase. It can therefore be difficult to work out what the correct form of a general conjecture should be, based on a finite amount of experimentation.

An example of this that follows from topics that we have already discussed is the field of definition of normalized cuspidal eigenforms for $\mathrm{SL}_2(\mathbf{Z})$. If we consider weights ≤ 26, then a large majority of these eigenforms have Fourier coefficients which are rational integers. However, the references given for Exercise 15 suggest that this is in fact completely atypical, and that in fact the characteristic polynomials of the Hecke operators are irreducible with Galois group as large as possible (S_n, the symmetric group acting on n letters, where n is the dimension of the space $S_k(\mathrm{SL}_2(\mathbf{Z}))$) (this is Maeda's Conjecture).

A more classical example from number theory is the number of primes of the form $4n+1$ less than x compared to the number of primes of the form $4n-1$ less than x. If we compute the first two thousand primes then we see that there are always more primes of the form $4n-1$ than primes of the form $4n+1$. However, it was proved by Littlewood in [Littlewood (1914)] in 1914 that there are infinitely many integers x such that this is reversed; there are more primes of the form $4n+1$ than primes of the form $4n-1$ less than x. This is a good example of something that could never be shown by computation, and indeed it would be hard to even come to the right conjecture by merely computing millions of primes. A gentle introduction to this subject is available at [Granville and Martin (2006)].

However, there are sometimes unusual circumstances where we can use computers to prove theorems for us. If, for instance, we know that we are dealing with a finite-dimensional vector space, or we know that certain types of relations must hold, then we can prove relations by finding a basis and then writing one element in terms of another. In [Gosper (1990)], the computer algebra package MACSYMA is used to prove combinatorial identities such as

$$\sum_{k=1}^{\infty} \frac{30k-11}{4(2k-1)k^3\binom{2k}{k}^2} = \zeta(3) \text{ and } \sum_{n=1}^{\infty} \frac{2^{-n}}{1+x^{2^{-n}}} = \frac{1}{\log x} + \frac{1}{1-x},$$

in the style of Ramanujan.

It can also happen that one has a formal structure in which genuine proofs can be given by computer. In [Wilf (2006)], a fascinating introduction to the world of generating functions is given; a method called the "Snake Oil Method" is introduced, which can give genuine although sometimes rather unilluminating proofs of combinatorial identities.

It should also be noted that computations can help to gather data to support conjecture; given enough data one can hope to find at least the shape of the correct result. There are books like [Rodriguez Villegas (2007)] which discuss the status of number theory as an "experimental science", in the words of Lehmer and Cassels. A very good example of this is given by the Birch–Swinnerton-Dyer Conjecture, as we mentioned earlier (see Section 20.5 of [Ireland and Rosen (1990)] for some background reading, and [Birch and Swinnerton-Dyer (1965)] for the original paper), where evidence gained from the EDSAC computer at the University of Cambridge helped to motivate the initial form of the conjecture. The Birch–Swinnerton-Dyer Conjecture is another of the Millennium Prize Problems, which means that there is a $\$1,000,000$ prize on offer for a proof of it.

7.5.6 *How hard is it to calculate Fourier coefficients of modular forms?*

The Fourier coefficients of modular forms are interesting to us for arithmetic reasons; it is useful to know how difficult it is to calculate them and what knowledge of the Fourier coefficients is equivalent to.

For instance, if p and q are prime numbers and we know the $(pq)^{\text{th}}$ Fourier coefficient of the normalized Eisenstein series E_k then we can factorize pq, and vice versa. This follows because

$$\sigma_{k-1}(pq) = 1 + p^{k-1} + q^{k-1} + (pq)^{k-1},$$

so we can extract $p^{k-1} + q^{k-1}$ from this by subtracting $1 + (pq)^{k-1}$. We then find the two roots of the polynomial $x^2 - (p^{k-1} + q^{k-1}) \cdot x + (pq)^{k-1}$, which are p^{k-1} and q^{k-1}. From these we can easily take $(k-1)^{\text{th}}$ roots, obtaining p and q. The converse is much easier; if we know p and q then we can clearly compute p^{k-1} and q^{k-1} and hence $\sigma_{k-1}(q)$. More generally, knowing the nth Fourier coefficient of E_k is equivalent to being able to factor n, in that given one there is a polynomial-time algorithm to find the other.

These are not strictly modular forms results, as they depend on arguments from the theory of arithmetic functions, but they do show us that computing the $(pq)^{\text{th}}$ coefficient of a modular form is at least as hard as factoring pq, because f could be a linear multiple of E_k.

In [Edixhoven *et al.* (2006)] it is proved that if p is a prime then one can compute $\tau(p)$ using a probabilistic algorithm with expected running time polynomial in p. This had long been suspected to be true. It is not known if there is a faster way of computing $\tau(n)$ than factoring n and then computing $\tau(p^m)$ for every $p^m || n$. Their research project aims to find a *deterministic* algorithm and extend this work to other modular forms with similar properties, such as the normalized cusp forms for $\mathrm{SL}_2(\mathbf{Z})$ of weights 16, 18, 20, 22 and 26.

7.6 Exercises

For some of these questions, you may have to look up commands in the MAGMA, SAGE or PARI documentation, which are all available online.

(1) Work out how large a collection of mathematical tables printed on paper would have to be in order to weigh two stones or two hundredweight (a British hundredweight is 112 pounds in weight).

(2) Research the growth of the largest prime number over time, and consider how computation will change in the future. You may find the Prime Pages ([Caldwell (2007)]) a useful resource for this question.
(3) Using a variety of algebra packages, find the factorization of $2^{67} - 1$, and see how long each takes to finish the computation.

7.6.1 MAGMA

(4) Think of several ways to check MacMahon's computation of $p(200)$, and implement them. Verify your computation with `NumberOfPartitions`. What happens if you try to compute this with `#Partitions(200)`?
(5) Use MAGMA to find congruences modulo p between Eisenstein series and cuspidal eigenforms of level 1. You may wish to proceed as follows:
   (a) Find spaces M of modular forms of dimension 2.
   (b) Find a basis of eigenforms for M to an appropriate precision.
   (c) Spot what the congruences must be by examining the first few coefficients of the eigenforms.
   (d) Construct a `PowerSeriesRing` over an appropriate `FiniteField` and use this to check that the difference d of the Eisenstein series and the cuspidal eigenform vanishes modulo p.
   (e) Find the Sturm bound for the weight and level and conclude that d must vanish completely.
(6) In this question, we will illustrate an example given in Section 7.5.3.
   (a) Choose a positive integer N.
   (b) Using the `Eisenstein` command, compute the coefficients of the Δ function up to precision $O(q^N)$. (You will need to define a `PowerSeriesRing` also).
   (c) Using the `DedekindEta` command, compute the coefficients of the Δ function up to precision $O(q^N)$.
   (d) Gather data on how long MAGMA takes to compute Δ in each of these two ways, using the `time` command or otherwise.
(7) Test the Shimura-Taniyama-Weil conjecture in the following way:
   (a) Compute the space of modular forms of level N and weight 2 for some N. Find the newforms f of that level which have coefficients in $\mathbf{Q}$.
   (b) Find the number of isogeny classes of elliptic curves E of conductor N.

(c) Show that there are pairs (E, f) such that the trace of E over $\mathbf{F}_p$ is equal to the pth coefficient of f (up to some reasonable bound).

(8) By considering spaces of modular forms of level 4 and even weight k, investigate how many representations of an integer there are as a sum of $2k$ squares. You may wish to consider Glaisher's approach in [Glaisher (1907a,b,c)].

(9) For a more open-ended challenge, consider how you would write a program that computed the Petersson scalar product $\langle f, g\rangle$ on spaces of modular forms in terms of $\langle f, f\rangle$ for f an eigenform.

(a) Let us assume that we are dealing with $M_k(\mathrm{SL}_2(\mathbf{Z}))$. First, check that at least one of f and g is a cusp form. Deal with the case when they are both non-cuspidal.

(b) Next, deal with the case when exactly one of the forms is non-cuspidal.

(c) Use the fact that $S_k(\mathrm{SL}_2(\mathbf{Z}))$ has a basis of eigenforms to write f and g as a sum of eigenforms.

(d) Finally, use the linearity and antilinearity properties of the scalar product to write $\langle f, g\rangle$ in terms of the scalar product acting on eigenforms.

How would you extend this to spaces of modular forms for other congruence subgroups?

(10) Use MAGMA to compute with spaces of half-integral weight modular forms.

(11) More generally, use MAGMA to compute spaces of modular forms, and familiarize yourself with its handling of modular forms. Use the `DirichletGroup` command to create Dirichlet characters and then use these to define spaces of modular forms $M_k(\Gamma_0(N), \chi)$ for some choices of N and χ.

(12) Investigate the "verbose" option in MAGMA. Enable verbose printing for the `Factorization` command, and see what techniques MAGMA uses to factorize an integer.

7.6.2 SAGE

(13) Investigate congruences between the two eigenforms in $S_2(\Gamma_0(37))$ using SAGE. Find other congruences.

(a) We first compute the two eigenforms using SAGE.

(b) Then we compute their Fourier expansions to a suitable precision; choose such a precision and justify it.
(c) Look at the difference of their Fourier expansions and see if there is a prime number which divides every term of the difference.

(14) Use SAGE to investigate the Ramanujan-Petersson conjecture; compare the size of coefficients of cuspidal eigenforms with those of Eisenstein series.
(a) Choose a level and a weight and compute a basis for the Eisenstein series, and compute some cuspidal modular forms.
(b) Can you see a difference between the cuspidal and non-cuspidal modular forms "by eye"? Are the coefficients of each genuinely different?
(c) Now choose a level and weight such that there are several Eisenstein series with nonzero constant coefficient. Choose a linear combination f of these which has no constant coefficient. How big are the coefficients of f?

(15) Investigate Maeda's Conjecture, that the characteristic polynomials of the Hecke operators T_p acting on $S_k(\mathrm{SL}_2(\mathbf{Z}))$ are irreducible over $\mathbf{Q}$.
(a) Look up the literature; Buzzard's paper [Buzzard (1996)], the paper of Conrey, Farmer and Wallace [Conrey *et al.* (2000)], and the paper of Farmer and James [Farmer and James (2002)]. Repeat the computations in these papers.
(b) Choose other weights k and primes p and check that the characteristic polynomials of the Hecke operators T_p acting on $S_k(\mathrm{SL}_2(\mathbf{Z}))$ are irreducible, by first computing the space of cuspforms and then factoring the characteristic polynomials.
(c) Show that the Galois group of the polynomials is always the full symmetric group. You may wish to use the techniques in Buzzard's paper to help with this, once the dimension gets large enough.
(d) Find out how long each of these computations takes, using the `time` command.

(16) As with MAGMA, investigate spaces of modular forms using SAGE; compare and contrast its handling of modular forms to that of MAGMA. Make yourself familiar with its capabilities. Investigate the built-in methods that are included with a space of modular forms.

(17) Investigate SAGE scripts, and write some yourself.

(18) Use the SAGE `plot` command to produce graphical output; for instance, replicate Figure 5.1. (The code used is given in the Hints and Solutions

section at the back of the book).

7.6.3 Pari

(19) Use Pari to compute η-products and η-quotients, and compare the Pari function `eta` with the Magma function `DedekindEta`. Can you construct a way to time these functions?

(20) Use the `plot`, `ploth` and `psploth` commands to produce plots with Pari. In particular, duplicate Figure 7.2. (The code can be found in the Hints and Solutions section at the back of the book).

(21) (This question comes from Section 8 of the Pari tutorial [Batut *et al.* (2000–2006)]). In this question we will see that strange things can happen when we attempt to do computations with floating-point numbers.

(a) Compute the GCD of `x^3-1` and `x^3 + x^2 + x - 3` using the `gcd` command.

(b) Now compute the gcd of `x^3-1.0` and `x^3 + x^2 + x - 3.`

(c) Is this an appropriate answer? Justify your reasoning.

(22) Use Pari to compute the first 10 nonzero Bernoulli numbers B_n as both rational numbers and real numbers.

7.6.4 Maple

(23) Download QSERIES and experiment with it. Compare and contrast Maple with Magma and Pari — which do you find easier to use for which sets of tasks?

(24) Choose a modular form $f \in M_k(\mathrm{SL}_2(\mathbf{Z}))$. Use the `solve` command to write f as a polynomial in E_4 and E_6 in the following way:

(a) Write down all of the $E_4^a \cdot E_6^b$ with $4a + 6b = k$. Compute their Fourier expansions to a suitable precision.

(b) Find a set of simultaneous expressions in the coefficients of the expansion

$$f = \sum_{\substack{4a+6b=k \\ a,b \text{ in } \mathbf{N}}} c_{a,b} \cdot E_4^a \cdot E_6^b$$

and solve them.

Appendix A

MAGMA code for classical modular forms

Here we present some examples of MAGMA intrinsics to illustrate the discussions of Section 7.2.2. To use these, save the intrinsics as a text file (called, say, `AppendixA.m`) in MAGMA's working directory, then use the `Attach` command to attach the intrinsics.

```
intrinsic ModularFormsDerivative(f::ModFrmElt) -> ModFrmElt
{This computes the \partial_k map from M_k to M_(k+2)}
n:=Dimension(Parent(f))+1;
PSR<q>:=PowerSeriesRing(BaseRing(Parent(f)),n);
E2:= 1-24*&+[q^i*&+Divisors(i): i in [1..n]];
fourierseries:=PSR!(12*q*Derivative(qExpansion(f,n))\
-E2*Weight(f)*qExpansion(f,n));
return ModularForms(Level(f),Weight(f)+2)!fourierseries;
end intrinsic;

intrinsic ModularFormsDerivative(f::ModFrmElt,M::ModFrm)
-> ModFrmElt
{This computes the \partial_k map from M_k to M_(k+2)}
n:=Dimension(Parent(f))+1;
PSR<q>:=PowerSeriesRing(BaseRing(Parent(f)),n);
E2:= 1-24*&+[q^i*&+Divisors(i): i in [1..n]];
fourierseries:=PSR!(12*q*Derivative(qExpansion(f,n))\
-E2*Weight(f)*qExpansion(f,n));
return M!fourierseries;
end intrinsic;
```

Appendix B

SAGE code for classical modular forms

We recall that SAGE uses Python syntax for programming. One of the interesting results of this is that whitespace is significant; it is used to delimit loops, and we see that the `for` loops in `E2q` are indented.

```
def E2q(n):
        PSR.<q>=PowerSeriesRing(QQ)
        temp=1
        for i in range(1,n):
                divs=divisors(i)
                for j in range(0,divs.__len__()):
                        temp=temp-24*divs[j]*q^i
        return temp

def ModularFormsDerivative(f,k,n):
        PSR.<q>=PowerSeriesRing(QQ)
        e2q=E2q(n)
        fourierseries=12*q*f.derivative()-e2q*k*f
        return fourierseries+O(q^(n+1))
```

Appendix B

SAGE code for classical modular forms

[illegible]

```
def E2(n):
    R.<q>=PowerSeriesRing(QQ)
    temp=1
    for i in range(1,n):
        d=[illegible]divisors(i)
        for [illegible] in range(0,[illegible]len(d)):
            temp=temp-24*d[illegible]*q^i
    return temp

def [illegible]Laplace[illegible](k,n):
    R.<q>=PowerSeriesRing(QQ)
    [illegible]=E2(n)
    [illegible]
    return [illegible]
```

Appendix C

Hints and answers to selected exercises

Chapter 2

Exercise 6: For the first part, show that every cusp of the form a/b with $(b, p) = 1$ is equivalent to 0 and that every cusp of the form a/b with $(b, p) = p$ is equivalent to ∞. For the second part, the cusps are all equivalent to one of $1/1$ (odd numerator, odd denominator), $0/1$ (even numerator, odd denominator) or ∞ (odd numerator, even denominator); show this by checking that the action of $\Gamma(2)$ preserves these three classes.
Exercise 8: In Figure C.1, we see a fundamental domain with a visible cusp at 0.
Exercise 11: the index is

$$N \prod_{p|N} \left(1 - \frac{1}{p}\right);$$

this can be shown by showing that the quotient is isomorphic to $(\mathbf{Z}/N\mathbf{Z})^\times$.
Exercise 13: if we do not impose the extra condition and just consider $\alpha\Gamma\alpha^{-1}$, then we may not have matrices with integer entries; consider $\alpha \left(\begin{smallmatrix} 1 & 0 \\ 1 & 1 \end{smallmatrix}\right) \alpha^{-1}$, where $\alpha = \left(\begin{smallmatrix} 4 & 0 \\ 0 & 1 \end{smallmatrix}\right)$.
Exercise 14: the reason that $\partial_{12}\Delta = 0$ is because ∂_k maps cusp forms to cusp forms, and $S_{14}(\mathrm{SL}_2(\mathbf{Z})) = 0$.
Exercise 17: by Proposition 2.9, we see that E_2 is holomorphic on $\mathcal{H}$ and at ∞, and because it has a Fourier expansion in terms of q we only need to check that $E_2^2 - \theta(E_2)$ transforms correctly under the action of $\left(\begin{smallmatrix} 0 & -1 \\ 1 & 0 \end{smallmatrix}\right)$ to show that it transforms correctly under all of $\mathrm{SL}_2(\mathbf{Z})$. This can be done directly by using the transformation relation from Proposition 2.9.
Exercise 18: the generalization of this result follows by explicit computation. For the second part, there is no reason in general that $\partial_k(f)$ should be a cusp form in general; it is only guaranteed to have a cusp at ∞.

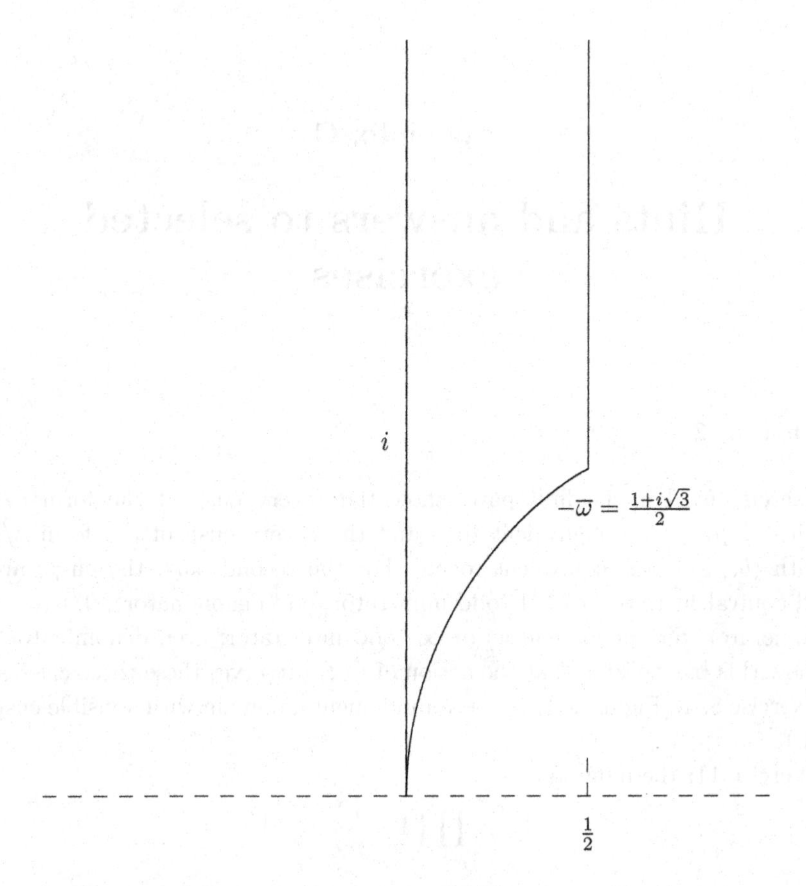

Fig. C.1 A fundamental domain for $\mathrm{SL}_2(\mathbf{Z})$ with a visible cusp at 0

Chapter 3

Exercise 1: this is a standard argument; we first prove the result for all modular forms of weight less than or equal to 12, fix an integer k, and write our modular form of weight k as the sum of a cusp form and a product $c \cdot E_4^a E_6^b$, and then use an induction argument on k.

Exercise 4: Let f be a nonzero cusp form. For the first part, take the modular form $f(q^N)$, for N arbitrarily large. For the second part, take the modular form $f(q)^N$, again with N arbitrarily large.

Chapter 4

Exercise 2: Consider, for example, the modular form $\Delta(q^2)$ as a modular form in $S_{12}(\Gamma_0(2))$.
Exercise 5: one way to tackle this is to consider spaces of cuspforms of dimension 2; we find that the eigenvalues of T_2 and T_3 acting on the space $S_{14}(\Gamma_0(2))$ are all even integers; in particular, the matrix $\left(\begin{smallmatrix} 1 & 0 \\ 0 & 1 \end{smallmatrix}\right)$ is not in the algebra generated by T_2 and T_3, but it is in the algebra which also has T_1 as a generator.
Exercise 6: The Hecke algebra $\mathbf{T}_{\mathbf{Z}}(S_2(\Gamma_0(37)))$ is

$$\{(a, b) : a, b \in \mathbf{Z} | \ a \equiv b \mod 2\}.$$

The Sturm bound in this case is 7.
Exercise 10: the Ramanujan-Petersson conjecture gives us a bound on the coefficients of a cusp form; the coefficients of E_2^* are much larger than this. This is possible because E_2^* is not a cusp form; it does vanish at ∞, but not at the other cusp 0. The congruence subgroup $\Gamma_0(p)$ has more than one cusp, so the fact that F vanishes at ∞ does not automatically make it a cusp form.
Exercise 2: We have seen that $\Gamma_0(2)$ has two cusps, so there can be two Eisenstein series which each vanish at one cusp. The modular forms E_4 and $(E_2^*)^2$ are both not cusp forms, and differ in their orders of vanishing at various points.
Exercise 11: if both f and g vanish at ∞, then their product will have a double zero at ∞; this will prevent it being an eigenform. Examples of eigenforms that are products of eigenforms are $\Delta_{16} = \Delta \cdot E_4$ and $E_{10} = E_6 \cdot E_4$.
Exercise 12: if we have a nonzero cusp form f of weight 1, then f^2 is a modular form of weight 2 whose Fourier expansion at ∞ begins $q^2 + O(q^3)$. This is not an eigenform; in particular, $T_2(f^2)$ will have Fourier expansion beginning $q + O(q^2)$. This gives us two linearly independent modular forms, and therefore the dimension of the space of weight 2 forms is at least 2.
Exercise 16: if the congruence subgroup Γ contains $-I$ then there can be no nonzero forms of odd weight.
Exercise 21: we can use the dimension formulae to deal with the even weight case, and the fact that there are no other odd characters modulo 2 or 4 to deal with the even weight case.

Chapter 5

Exercise 2: the level here is $\Gamma_0(8)$; the dimension of the space of Eisenstein series is 3, and there are no cusp forms. A basis for the space of modular forms is given by $\{E_2(q) - 2E_2(q^2), E_2(q) - 4E_2(q^4), E_2(q) - 8E_2(q^8)\}$.
Exercise 4: the dimension here is 2 and the Sturm bound is 1.
Exercise 5: a set of coset representatives for $\Gamma(2)$ in $\mathrm{PSL}_2(\mathbf{Z})$ is given by $\{\left(\begin{smallmatrix}1&0\\0&1\end{smallmatrix}\right), \left(\begin{smallmatrix}0&1\\-1&1\end{smallmatrix}\right), \left(\begin{smallmatrix}-1&1\\-1&0\end{smallmatrix}\right)\}$. The function λ is not a modular function for $\Gamma_1(2)$ because it does not have a Fourier expansion in terms of q at the cusp ∞, and it is not a modular form because it has weight 0 and is not a constant.
Exercise 6: the set $\{E_4(q), E_4(q^2), E_4(q^4)\}$ is a basis for $M_4(\Gamma_0(4))$. The space of cusp forms has dimension 0.
Exercise 9: the identities are

$$j = \frac{(j_2 + 256)^3}{j_2^2} \text{ and } j_2 = \frac{j_4^2}{j_4 + 16}.$$

Exercise 10 We give an example for $p = 2$. The formula for $j_2(q)$ is

$$j_2(q) = 21493760 \cdot \Phi(q) + 19730006016 \cdot \Phi(q)^2 + 3298534883328 \cdot \Phi(q)^3 + 140737488355328 \cdot \Phi(q)^4.$$

We also need to check that $j_p(q)$ is a modular function itself. This follows because if f is a modular function of weight k for $\mathrm{SL}_2(\mathbf{Z})$, then

$$\frac{1}{p}\sum_{i=0}^{p-1} f\left(\frac{z+i}{p}\right)$$

is a modular function of weight k for $\Gamma_0(p)$.
Exercise 11: The nine negative d for which $\mathbf{Q}(\sqrt{d})$ has class number one are: $\{-1, -2, -3, -7, -11, -19, -43, -67, -163\}$. Each of these has an "almost integer" $e^{\pi d}$ associated to it.
Exercise 12: The identity of weight 12 modular forms is

$$432000\Delta = 691E_4^3 - 691E_{12}.$$

From this, we can reduce everything modulo 691; the terms in $691E_4^3$ all vanish, as does the constant term of $691E_{12}$. The fact that $432000 \equiv -65520$ modulo 691 will then prove the result.
Exercise 15: for the penultimate part, we note that if θ^4 contributes any zeroes to the modular form G_6, then θ^2 will also contribute those zeroes.
Exercise 17: A suitable elliptic curve of conductor 32 is $E : y^2 = x^3 + 4x$.
Exercise 20: Consider $q = (1\ i)$, and show that $q^T A q = 0$. For the final part, note that the construction still works when $k \equiv 3$ modulo 4. Evaluate it to see what happens.

Chapter 6

Exercise 4: the Eisenstein series can be taken to be E_{10} and the Sturm bound for $S_{12}(\Gamma_0(11))$ is 12.
Exercise 5: one way to do this is to write down a linear combination of these Eisenstein series and use the fact that the zeroes of E_4 and E_6 are distinct.
Exercise 7: we can use dimensional analysis to check that if $E(q)$ is a modular form then it has the correct weight by counting the number of variables, and using the results we have on theta functions to show that it is a modular form for $\Gamma_1(12)$. Using more sophisticated techniques one can actually show that $U_2(E(q))$ is the Fourier expansion of a modular form for $\Gamma_1(3)$.

Chapter 7

Exercise 4: one could use the recurrence relation for $p(n)$ given in (5.43), for instance. Asking for `#Partitions(200)` is likely to take a very long time.
Exercise 18: the code used here was

```
E=EllipticCurve([0,17])
G=plot(E,xmin=-4,xmax=5,rgbcolor=hue(0.5))
P1andP2=line([[-2,-3],[4,9]], rgbcolor=hue(0.3))
P4vertical=line([[4,9],[4,-9]], rgbcolor=hue(0.7))
l1=text('P1',(-2.25,-3.25))
l2=text('P2',(1.75,5.25))
l3=text('P1 + P2',(3.5,-9.25))
H=(G + P1andP2 + P4vertical + l1 + l2 + l3)
H.axes(false)
H.save("ellcurve.eps")
```

Exercise 20: the code used was

```
psploth(X=-10,10,abs(X))
```

Exercise 21: The problem here is that we are trying to take greatest common divisors over $\mathbf{R}$ (the use of 1.0) rather than $\mathbf{Z}$.

Bibliography

Agrawal, M., Kayal, N. and Saxena, N. (2004). PRIMES is in P, *Ann. of Math. (2)* **160**, 2, pp. 781–793.

Ahlgren, S. and Barcau, M. (2007). Congruences for modular forms of weights two and four. *J. Number Theory* **126**, 2, pp. 193–199.

Andrews, G. E. (1976). *The Theory of Partitions* (Addison-Wesley).

Andrews, G. E. (1983). Euler's pentagonal number theorem, *Math. Mag* **56**, 5, pp. 279–284.

Apostol, T. M. (1976). *Modular functions and Dirichlet series in number theory* (Springer-Verlag, New York), Graduate Texts in Mathematics, No. 41.

Armitage, J. V. and Eberlein, W. F. (2006). *Elliptic functions, London Mathematical Society Student Texts*, Vol. 67 (Cambridge University Press, Cambridge), ISBN 978-0-521-78078-0; 978-0-521-78563-1; 0-521-78563-4.

Atkin, A. O. L. (1968). Feasible Computability, in *Some Research Applications of the Computer* (SERC, for the Atlas Computer Laboratory), pp. 3–6.

Atkin, A. O. L. and Garvan, F. G. (2003). Relations between the ranks and cranks of partitions, *Ramanujan J.* **7**, 1-3, pp. 343–366, rankin memorial issues.

Atkin, A. O. L. and Lehner, J. (1970). Hecke operators on $\Gamma_0(m)$, *Math. Ann.* **185**, pp. 134–160.

Atkin, A. O. L. and Morain, F. (1993). Elliptic curves and primality proving, *Math. Comp.* **61**, 203, pp. 29–68.

Baker, A. (1990). *Transcendental number theory*, 2nd edition, Cambridge Mathematical Library (Cambridge University Press, Cambridge), ISBN 0-521-39791-X.

Batut, C., Belabas, K., Bernardi, D., Cohen, H. and Olivier, M. (2000–2006). A Tutorial for Pari/GP, version 2.3.1, `http://pari.math.u-bordeaux.fr/pub/pari/manuals/2.3.1/`.

Berrizbeitia, P. (2005). Sharpening "PRIMES is in P" for a large family of numbers, *Math. Comp.* **74**, 252, pp. 2043–2059 (electronic).

Birch, B. (1998). Atkin and the Atlas Lab, in *Computational perspectives on number theory (Chicago, IL, 1995), AMS/IP Stud. Adv. Math.*, Vol. 7 (Amer. Math. Soc., Providence, RI), pp. 13–20.

Birch, B. J. and Kuyk, W. (eds.) (1975). *Modular functions of one variable. IV*

(Springer-Verlag, Berlin), Lecture Notes in Mathematics, Vol. 476.

Birch, B. J. and Swinnerton-Dyer, H. P. F. (1965). Notes on elliptic curves. II, *J. Reine Angew. Math.* **218**, pp. 79–108.

Borcherds, R. E. (1992). Monstrous moonshine and monstrous Lie superalgebras, *Invent. Math.* **109**, 2, pp. 405–444.

Borwein, J. M. and Borwein, P. B. (1987). *Pi and the AGM — A Study in Analytic Number Theory and Computational Complexity* (Wiley, N.Y).

Borwein, J. M., Borwein, P. B. and Bailey, D. H. (1989). Ramanujan, Modular Equations, and Approximations to Pi or How to Compute One Billion Digits of Pi, *The American Mathematical Monthly* **96**, 3, pp. 201–219.

Bosma, W. and Cannon, J. (eds.) (2006). *Discovering mathematics with Magma — Reducing the Abstract to the Concrete, Algorithms and Computation in Mathematics*, Vol. 19 (Springer-Verlag, Berlin), ISBN 978-3-540-37632-3; 3-540-37632-1.

Bosma, W., Cannon, J. and Playoust, C. (1997). The Magma algebra system I: The user language, *J. Symb. Comp.* **24**, 3–4, pp. 235–265, available from `http://magma.maths.usyd.edu.au`.

Bosman, J. (2007). On the computation of Galois representations associated to level one modular forms, URL `http://www.citebase.org/abstract?id=oai:arXiv.org:0710.1237`.

Breuil, C., Conrad, B., Diamond, F. and Taylor, R. (2001). On the modularity of elliptic curves over **Q**: wild 3-adic exercises, *J. Amer. Math. Soc.* **14**, 4, pp. 843–939 (electronic).

Brillhart, J., Lehmer, D. H., Selfridge, J. L., Tuckerman, B. and Wagstaff, S. S., Jr. (2002). *Factorizations of* $b^n \pm 1$, $b = 2, 3, 5, 6, 7, 10, 11, 12$ *up to high powers*, *Contemporary Mathematics*, Vol. 22, 3rd edition (American Mathematical Society, Providence, RI), ISBN 0-8218-5078-4, available at `http://www.ams.org/online_bks/conm22/`.

Bringmann, K. and Ono, K. (2007). Lifting cusp forms to Maass forms with an application to partitions, *Proceedings of the National Academy of Sciences of the USA* **104**, 10, pp. 3725–3731.

Bryan, J. and Leung, N. C. (2000). The enumerative geometry of $K3$ surfaces and modular forms, *J. Amer. Math. Soc.* **13**, 2, pp. 371–410 (electronic).

Buhler, J. P. (2001). Elliptic curves, modular forms, and applications, in *Arithmetic algebraic geometry (Park City, UT, 1999), IAS/Park City Math. Ser.*, Vol. 9 (Amer. Math. Soc., Providence, RI), pp. 5–81.

Bump, D. (1997). *Automorphic forms and representations*, *Cambridge Studies in Advanced Mathematics*, Vol. 55 (Cambridge University Press, Cambridge), ISBN 0-521-55098-X.

Bump, D., Cogdell, J. W., de Shalit, E., Gaitsgory, E., D. Kowalski and Kudla, S. S. (2003). *An introduction to the Langlands program* (Birkhäuser Boston Inc.), lectures presented at the Hebrew University of Jerusalem, Jerusalem, March 12–16, 2001. Edited by J. Bernstein and S. Gelbart.

Buzzard, K. (1996). On the eigenvalues of the Hecke operator T_2, *J. Number Theory* **57**, 1, pp. 130–132.

Buzzard, K. (2005). Questions about slopes of modular forms, *Astérisque* **298**, pp. 1–15, automorphic forms. I.

Buzzard, K. and Calegari, F. (2004). A counterexample to the Gouvêa-Mazur conjecture, *C. R. Math. Acad. Sci. Paris* **338**, 10, pp. 751–753.

Caldwell, C. (2007). The Prime Pages, `http://primes.utm.edu`.

Calegari, F. and Stein, W. A. (2004). Conjectures about discriminants of Hecke algebras of prime level. Buell, Duncan (ed.), Algorithmic number theory. 6th international symposium, ANTS-VI, Burlington, VT, USA, June 13–18, 2004. Proceedings. Berlin: Springer. Lecture Notes in Computer Science 3076, 140-152 (2004).

Campbell-Kelly, M., Croarken, M., Flood, R. and Robson, E. (eds.) (2003). *The history of mathematical tables — From Sumer to spreadsheets* (Oxford University Press, Oxford), ISBN 0-19-850841-7.

Cannon, J. J. (1984). An introduction to the group theory language, Cayley, in *Computational group theory (Durham, 1982)* (Academic Press, London), pp. 145–183.

Cassels, J. W. S. (1991). *Lectures on elliptic curves, London Mathematical Society Student Texts*, Vol. 24 (Cambridge University Press, Cambridge), ISBN 0-521-41517-9; 0-521-42530-1.

Clarke, A. C. (1962). *Profiles of the Future*, chap. Hazards of Prophecy: The Failure of Imagination (Harper and Row).

Clay, L. (2005). *Some Conjectures About the Slopes of Modular Forms*, Ph.D. thesis, Northwestern.

Cohen, H. (1993). *A course in computational algebraic number theory, Graduate Texts in Mathematics*, Vol. 138 (Springer-Verlag, Berlin), ISBN 3-540-55640-0.

Cohen, H. (2000). *Advanced topics in computational number theory, Graduate Texts in Mathematics*, Vol. 193 (Springer-Verlag, New York), ISBN 0-387-98727-4.

Cohen, H. and Oesterlé, J. (1977). Dimensions des espaces de formes modulaires, *Lecture Notes in Mathematics* **627**, pp. 69–78.

Cole, F. N. (1903). On the factoring of large numbers. *Bulletin of the American Mathematical Society* **2**, 10, pp. 184–187.

Coleman, R., Stevens, G. and Teitelbaum, J. (1998). Numerical experiments on families of p-adic modular forms, *AMS/IP Studies in Advanced Mathematics* **7**, pp. 143–158.

Conrad, B., Diamond, F. and Taylor, R. (1999). Modularity of certain potentially Barsotti-Tate Galois representations, *J. Amer. Math. Soc.* **12**, 2, pp. 521–567.

Conrey, J. B., Farmer, D. W. and Wallace, P. J. (2000). Factoring Hecke polynomials modulo a prime, *Pacific J. Math.* **196**, 1, pp. 123–130.

Conway, J. H. and Norton, S. P. (1979). Monstrous moonshine, *Bull. London Math. Soc.* **11**, 3, pp. 308–339.

Coogan, G. H. (2005). Congruences concerning values of the modular functions j_n, *Ramanujan J.* **10**, 1, pp. 43–50.

Cornell, G., Silverman, J. H. and Stevens, G. (eds.) (1997). *Modular forms and Fermat's last theorem* (Springer-Verlag, New York), ISBN 0-387-94609-8; 0-

387-98998-6, papers from the Instructional Conference on Number Theory and Arithmetic Geometry held at Boston University, Boston, MA, August 9–18, 1995.

Cox, D. A. (1989). *Primes of the form* $x^2 + ny^2$, A Wiley-Interscience Publication (John Wiley & Sons Inc., New York), ISBN 0-471-50654-0; 0-471-19079-9, fermat, class field theory and complex multiplication.

Crandall, R. and Pomerance, C. (2005). *Prime Numbers — A Computational Perspective*, 2nd edition (Springer, New York), ISBN 978-0-387-25282-7; 0-387-25282-7.

Cremona, J. E. (1997). *Algorithms for modular elliptic curves*, 2nd edition (Cambridge University Press, Cambridge), ISBN 0-521-59820-6.

Croarken, M. and Campbell-Kelly, M. (2000). Beautiful Numbers: The Rise and Decline of the British Association Mathematical Tables Committee, 1871–1965, *IEEE Ann. Hist. Comput.* **22**, 4, pp. 44–61, doi:http://dx.doi.org/10.1109/85.887989.

Darmon, H. (1999). A proof of the full Shimura-Taniyama-Weil conjecture is announced, *Notices Amer. Math. Soc.* **46**, 11, pp. 1397–1401.

David, K. (1987). The Teaching of Mathematics: Using Commutators to Prove A_5 is Simple, *Amer. Math. Monthly* **94**, 8, pp. 775–776.

Deligne, P. (1973). Formes modulaires et représentations de GL(2), in *Modular functions of one variable, II (Proc. Internat. Summer School, Univ. Antwerp, Antwerp, 1972)* (Springer, Berlin), pp. 55–105. Lecture Notes in Math., Vol. 349.

Deligne, P. (1974). La conjecture de Weil. I, *Inst. Hautes Études Sci. Publ. Math.* **43**, pp. 273–307.

Deligne, P. and Kuyk, W. (eds.) (1973). *Modular functions of one variable. II* (Springer-Verlag, Berlin), Lecture Notes in Mathematics, Vol. 349.

Deligne, P. and Serre, J.-P. (1974). Formes modulaires de poids 1, *Ann. Sci. École Norm. Sup. (4)* **7**, pp. 507–530 (1975).

Devlin, K. (2002). *The millennium problems* (Basic Books, New York), ISBN 0-465-01729-0; 0-465-01730-4, the seven greatest unsolved mathematical puzzles of our time.

Diamond, F. (1996). On deformation rings and Hecke rings, *Ann. of Math. (2)* **144**, 1, pp. 137–166.

Diamond, F. and Im, J. (1995). Modular Forms and Modular Curves, in *Seminar on Fermat's Last Theorem (Toronto, ON, 1993–1994)*, pp. 39–133.

Diamond, F. and Shurman, J. (2005). *A first course in modular forms*, *Graduate Texts in Mathematics*, Vol. 228 (Springer-Verlag, New York), ISBN 0-387-23229-X.

Dickson, L. E. (1966). *History of the theory of numbers. Vol. III: Quadratic and higher forms.*, With a chapter on the class number by G. H. Cresse (Chelsea Publishing Co., New York).

Edixhoven, B. (2006). Comparison of integral structures on spaces of modular forms of weight two, and computation of spaces of forms mod 2 of weight one, *J. Inst. Math. Jussieu* **5**, 1, pp. 1–34, with appendix A (in French) by Jean-François Mestre and appendix B by Gabor Wiese.

Edixhoven, B., Couveignes, J.-M., de Jong, R., Merkl, F. and Bosman, J. (2006). On the computation of coefficients of a modular form, Math.NT/0605244.

Emmons, B. A. (2005). Products of Hecke eigenforms, *J. Number Theory* **115**, 2, pp. 381–393.

Farmer, D. W. and James, K. (2002). The irreducibility of some level 1 Hecke polynomials, *Math. Comp.* **71**, 239, pp. 1263–1270 (electronic).

Ford, L. R. (1957). *Automorphic Functions*, 2nd edition (Chelsea Pub. Co., New York), reprinted from first edition, 1929.

Frey, G. (1986). Links between stable elliptic curves and certain Diophantine equations, *Ann. Univ. Sarav. Ser. Math.* **1**, 1, pp. iv+40.

Gannon, T. (2006). Monstrous moonshine: the first twenty-five years, *Bull. London Math. Soc.* **38**, 1, pp. 1–33.

Garvan, F. G. (1998). The MAPLE QSERIES Package, Available at `http://www.math.ufl.edu/~frank/qmaple/qmaple.html`.

Gauss, C. F. (1986). *Disquisitiones arithmeticae* (Springer-Verlag, New York), ISBN 0-387-96254-9, translated and with a preface by Arthur A. Clarke, Revised by William C. Waterhouse, Cornelius Greither and A. W. Grootendorst and with a preface by Waterhouse.

Glaisher, J. W. L. (1907a). On the representation of a number as the sum of eighteen squares, *Quarterly Journal of Pure and Applied Mathematics* **38**, pp. 289–352.

Glaisher, J. W. L. (1907b). On the representation of a number as the sum of fourteen and sixteen squares, *Quarterly Journal of Pure and Applied Mathematics* **38**, pp. 178–236.

Glaisher, J. W. L. (1907c). On the representation of a number as the sum of two, four, six, eight, ten or twelve squares, *Quarterly Journal of Pure and Applied Mathematics* **38**, pp. 1–62.

Gordon, B. and Robins, S. (1995). Lacunarity of Dedekind η-products. *Glasg. Math. J.* **37**, 1, pp. 1–14.

Gorenstein, D., Lyons, R. and Solomon, R. (1994). *The classification of the finite simple groups, Mathematical Surveys and Monographs*, Vol. 40 (American Mathematical Society, Providence, RI), ISBN 0-8218-0334-4.

Gosper, W. (1990). Strip mining in the abandoned orefields of nineteenth century mathematics, in *Computers in mathematics (Stanford, CA, 1986)*, *Lecture Notes in Pure and Appl. Math.*, Vol. 125 (Dekker, New York), pp. 261–284.

Gouvêa, F. and Mazur, B. (1992). Families of Modular Eigenforms, *Math. Comp.* **58**, pp. 793–805.

Gouvêa, F. Q. (1997). *p-adic numbers*, 2nd edition, Universitext (Springer-Verlag, Berlin), ISBN 3-540-62911-4, an introduction.

Granville, A. and Martin, G. (2006). Prime number races, *Amer. Math. Monthly* **113**, 1, pp. 1–33.

Green, B. and Tao, T. (2008). The primes contain arbitrarily long arithmetic progressions, *Annals of Mathematics* .

Grosswald, E. (1985). *Representations of integers as sums of squares* (Springer-Verlag, New York), ISBN 0-387-96126-7.

Grotendorst, J. (1991). A Maple package for transforming series, sequences and functions, *Comput. Phys. Comm.* **67**, 2, pp. 325–342.

Gupta, H., Cheema, M. S., Mehta, A. and Gupta, O. P. (1960). *Representations of primes by quadratic forms: displaying solutions of the Diophantine equation* $kp = a^2 + Db^2$. *Part I:* $D = 5, 6, 10,$ *and* 13, Royal Society Mathematical Tables, Vol. 5. (Cambridge University Press, New York).

Guy, R. K. (1988). The strong law of small numbers, *Amer. Math. Monthly* **95**, 8, pp. 697–712.

Hardy, G. H. (1920). On the representation of a number as the sum of any number of squares, and in particular of five, *Trans. Amer. Math. Soc.* **21**, 3, pp. 255–284.

Hardy, G. H. and Ramanujan, S. (1917). Asymptotic formulae in combinatory analysis. *Proceedings of the London Mathematical Society (2)* **17**, pp. 75–115.

Hardy, G. H. and Ramanujan, S. (1918). On the Coefficients in the Expansions of Certain Modular Functions, *Proceedings of the Royal Society of London. Series A* **95**, 667, pp. 144–155.

Hardy, G. H. and Wright, E. M. (1979). *An introduction to the theory of numbers*, 5th edition (The Clarendon Press Oxford University Press, New York), ISBN 0-19-853170-2; 0-19-853171-0.

Hartshorne, R. (1977). *Algebraic geometry* (Springer-Verlag, New York), ISBN 0-387-90244-9, graduate Texts in Mathematics, No. 52.

Heck, A. (1993). *Introduction to Maple* (Springer-Verlag, New York), ISBN 0-387-97662-0.

Hellegouarch, Y. (1974/75). Points d'ordre $2p^h$ sur les courbes elliptiques, *Acta Arith.* **26**, 3, pp. 253–263.

Hellegouarch, Y. (2002). *Invitation to the mathematics of Fermat-Wiles* (Academic Press Inc., San Diego, CA), ISBN 0-12-339251-9, translated from the second (2001) French edition by Leila Schneps.

Hermite, C. (1859). Sur la théorie des équations modulaires, *Comptes Rendus Acad. Sci. Paris* **49**, pp. 16–24, 110–118, and 141–144.

Hida, H. (2006). *Hilbert modular forms and Iwasawa theory*, Oxford Mathematical Monographs (The Clarendon Press Oxford University Press, Oxford), ISBN 978-0-19-857102-5; 0-19-857102-X.

Hsu, T. (1996). Identifying congruence subgroups of the modular group, *Proc. Amer. Math. Soc.* **124**, 5, pp. 1351–1359.

Hunt, B. R., Lipsman, R. L. and Rosenberg, J. M. (2006). *A guide to MATLAB® for beginners and experienced users, Updated for MATLAB® 7 and Simulink® 6*, 2nd edition (Cambridge University Press, Cambridge), ISBN 978-0-521-61565-5; 0-521-61565-8.

Husemöller, D. (2004). *Elliptic curves, Graduate Texts in Mathematics*, Vol. 111, 2nd edition (Springer-Verlag, New York), ISBN 0-387-95490-2, with appendices by Otto Forster, Ruth Lawrence and Stefan Theisen.

Ireland, K. and Rosen, M. (1990). *A classical introduction to modern number theory, Graduate Texts in Mathematics*, Vol. 84, 2nd edition (Springer-Verlag, New York), ISBN 0-387-97329-X.

Jarvis, F. and Meekin, P. (2004). The Fermat equation over $\mathbb{Q}(\sqrt{2})$, *J. Number*

Theory **109**, 1, pp. 182–196.

Jensen, K. L. (1915). Om talteoretiske Egenskaber ved de *Bernoulliske* Tal. *Nyt Tidsskr. for Math.* **26**, pp. 73–83.

Jochnowitz, N. (1982). Congruences between systems of eigenvalues of modular forms, *Trans. Amer. Math. Soc.* **270**, 1, pp. 269–285.

Jones, G. A. (1986). Congruence and noncongruence subgroups of the modular group: a survey, in *Proceedings of groups—St. Andrews 1985, London Math. Soc. Lecture Note Ser.*, Vol. 121 (Cambridge Univ. Press, Cambridge), pp. 223–234.

Jorgenson, J. and Krantz, S. G. (2006). Serge Lang, 1927–2005, *Notices of the American Mathematical Society* **53**, 5, pp. 536–553.

Kaneko, M. and Zagier, D. (1995). A generalized Jacobi theta function and quasimodular forms, in *The moduli space of curves (Texel Island, 1994), Progr. Math.*, Vol. 129 (Birkhäuser Boston, Boston, MA), pp. 165–172.

Katz, N. M. (1973). p-adic properties of modular schemes and modular forms, in *Modular functions of one variable, III (Proc. Internat. Summer School, Univ. Antwerp, Antwerp, 1972)* (Springer, Berlin), pp. 69–190. Lecture Notes in Mathematics, Vol. 350.

Khare, C. (2006a). Serre's modularity conjecture: a survey of the level one case, To appear in Proceedings of the LMS Durham conference.

Khare, C. (2006b). Serre's modularity conjecture: the level one case, *Duke Math. J.* **134**, 3, pp. 557–589.

Khare, C. and Wintenberger, J.-P. (2006a). Serre's modularity conjecture (i), Available at `http://www.math.utah.edu/~shekhar/papers.html`.

Khare, C. and Wintenberger, J.-P. (2006b). Serre's modularity conjecture (ii), Available at `http://www.math.utah.edu/~shekhar/papers.html`.

Kilford, L. J. P. (2002). Some non-Gorenstein Hecke algebras attached to spaces of modular forms, *Journal of Number Theory* **97**, pp. 157–164.

Kilford, L. J. P. and Wiese, G. (2008). On the failure of the Gorenstein property for Hecke algebras of prime weight, Math.NT/0612317. To appear in *Experimental Mathematics*.

Kisin, M. (2008). Modularity of 2-adic Barsotti-Tate representations, Available at `http://www.math.uchicago.edu/~kisin/preprints.html`.

Klingen, H. (1990) Introductory lectures on Siegel modular forms, Cambridge Studies in Advanced Mathematics, 20. Cambridge University Press, Cambridge.

Knapp, A. W. (1992). *Elliptic curves, Mathematical Notes*, Vol. 40 (Princeton University Press, Princeton, NJ), ISBN 0-691-08559-5.

Knapp, A. W. (1997). Introduction to the Langlands program, in *Representation theory and automorphic forms (Edinburgh, 1996), Proc. Sympos. Pure Math.*, Vol. 61 (Amer. Math. Soc.), pp. 245–302.

Knopp, M. I. (1970). *Modular Functions in Analytic Number Theory* (Markham).

Knuth, D. E. (1969a). *The art of computer programming. Vol. 2: Seminumerical algorithms* (Addison-Wesley Publishing Co., Reading, Mass.-London-Don Mills, Ont).

Knuth, D. E. (1969b). *The art of computer programming. Volume 1: Fundamental algorithms* (Addison-Wesley).

Knuth, D. E. (1974). Structured programming with goto statements, *ACM Comput. Surv.* **6**, 4, pp. 261–301.

Koblitz, N. (1993). *Introduction to Elliptic Curves and Modular Forms* (Springer, New York).

Krieg, A. (1990). Hecke algebras. *Mem. Am. Math. Soc.* **435**, p. 158 p.

Kurth, C. (2007). Farey Symbol Functions, Available from `http://www.public.iastate.edu/~kurthc/research/index.html`.

Kuyk, W. (ed.) (1973). *Modular functions of one variable. I* (Springer-Verlag, Berlin), lecture Notes in Mathematics, Vol. 320.

Kuyk, W. and Serre, J.-P. (eds.) (1973). *Modular functions of one variable. III* (Springer-Verlag, Berlin), Lecture Notes in Mathematics, Vol. 350.

Lamport, L. (1994). *LaTeX: a document preparation system: user's guide and reference manual*, Second edition (Addison-Wesley, Reading, Massachusetts, USA).

Lang, S. (1976). *Introduction to modular forms* (Springer-Verlag, Berlin), Grundlehren der mathematischen Wissenschaften, No. 222.

Lang, S. (1978). *Elliptic curves: Diophantine analysis, Grundlehren der Mathematischen Wissenschaften [Fundamental Principles of Mathematical Sciences]*, Vol. 231 (Springer-Verlag, Berlin), ISBN 3-540-08489-4.

Lansky, J. and Pollack, D. (2002). Hecke algebras and automorphic forms. *Compos. Math.* **130**, 1, pp. 21–48.

Lehmer, D. H. (1928). The Mechanical Combination of Linear Forms, *Amer. Math. Monthly* **35**, 3, pp. 114–121.

Lehmer, D. H. (1947). The vanishing of Ramanujan's function $\tau(n)$, *Duke Math. J.* **14**, pp. 429–433.

Lehner, J. (1949a). Divisibility properties of the Fourier coefficients of the modular invariant $j(\tau)$. *Am. J. Math.* **71**, pp. 136–148.

Lehner, J. (1949b). Further congruence properties of the Fourier coefficients of the modular invariant $j(\tau)$. *Am. J. Math.* **71**, pp. 373–386.

Li, W. C. W. (1975). Newforms and functional equations, *Math. Ann.* **212**, pp. 285–315.

Li, W.-C. W., Long, L. and Yang, Z. (2005). Modular forms for noncongruence subgroups, *Q. J. Pure Appl. Math.* **1**, 1, pp. 205–221.

Ligozat, G. (1975). *Courbes modulaires de genre* 1 (Société Mathématique de France, Paris), Bull. Soc. Math. France, Mém. 43, Supplément au Bull. Soc. Math. France Tome 103, no. 3.

Littlewood, J. E. (1914). Distribution des nombres premiers, *C. R. Acad. Sci. Paris* **158**, pp. 1869–1872.

Littlewood, J. E. (1986). *Littlewood's Miscellany* (Cambridge University Press).

Lutz, M. (2003). *Learning Python* (O'Reilly & Associates, Inc., Sebastopol, CA, USA), ISBN 0596002815.

Mazur, B. (1978). Rational isogenies of prime degree (with an appendix by D. Goldfeld), *Invent. Math.* **44**, 2, pp. 129–162.

McMurdy, K. (2001). *A Splitting Criterion for Galois Representations Associated to Exceptional Modular Forms*, Ph.D. thesis, U.C. Berkeley, available at http://phobos.ramapo.edu/~kmcmurdy/research/.

Milne, J. (1997a). Elliptic Curves, Online notes, available at http://www.jmilne.org/math/CourseNotes/math679.html.

Milne, J. (1997b). Modular Functions and Modular Forms, Online notes, available at http://www.jmilne.org/math/CourseNotes/math678.html.

Milne, J. (2006). *Elliptic Curves* (BookSurge Publishing).

Miyake, T. (1971). On automorphic forms on GL_2 and Hecke operators, *Ann. of Math. (2)* **94**, pp. 174–189.

Miyake, T. (2006). *Modular forms*, English edition, Springer Monographs in Mathematics (Springer-Verlag, Berlin), ISBN 978-3-540-29592-1; 3-540-29592-5.

Mordell, L. J. (1918). On the representations of numbers as a sum of 2^r squares. *Quart. J.* **48**, pp. 93–104.

Moreno, C. J. and Wagstaff, S. S., Jr. (2006). *Sums of squares of integers*, Discrete Mathematics and its Applications (Boca Raton) (Chapman & Hall/CRC, Boca Raton, FL), ISBN 978-1-58488-456-9; 1-58488-456-8.

Murty, M. R. (1993). A motivated introduction to the Langlands program, in *Advances in number theory (Kingston, ON, 1991)*, Oxford Sci. Publ. (Oxford University Press, New York, NY), pp. 37–66.

Murty, V. (1987). Lacunarity of modular forms. *J. Indian Math. Soc., New Ser.* **52**, pp. 127–146.

Newman, M. (1958). Congruences for the coefficients of modular forms and for the coefficients of $j(\tau)$, *Proc. Amer. Math. Soc.* **9**, pp. 609–612.

Odlyzko, A. M. and te Riele, H. J. J. (1985). Disproof of the Mertens Conjecture, *Journal fü die reine und angewandte Mathematik* **357**, pp. 138–160.

Ogg, A. (1969). *Modular forms and Dirichlet series* (W. A. Benjamin, Inc., New York-Amsterdam).

Ono, K. (1998). Gordon's ε-conjecture on the lacunarity of modular forms. *C. R. Math. Acad. Sci., Soc. R. Can.* **20**, 4, pp. 103–107.

Ono, K. (2004). *The web of modularity: arithmetic of the coefficients of modular forms and q-series, CBMS Regional Conference Series in Mathematics*, Vol. 102 (Published for the Conference Board of the Mathematical Sciences, Washington, DC), ISBN 0-8218-3368-5.

Pari (2005). *PARI/GP, version* 2.1.7, The PARI Group, Bordeaux, available from http://pari.math.u-bordeaux.fr/.

Petersson, H. (1932). Über die Entwicklungskoeffizienten der automorphen formen, *Acta Mathematica* **58**, pp. 169–215.

Press, W. H., Teukolsky, S. A., Vetterling, W. T. and Flannery, B. P. (2002). *Numerical recipes in C++ — The art of scientific computing* (Cambridge University Press, Cambridge), ISBN 0-521-75033-4.

Ramanujan, S. (1913–1914). Modular equations and approximations to π, *Quart. J. Pure Appl. Math.* **45**, pp. 350–372.

Ramanujan, S. (1916). On certain arithmetical functions, *Transactions of the Cambridge Philosophical Society* **22**, 9, pp. 159–184.

Rankin, R. A. (1956). The construction of automorphic forms from the derivatives of a given form, *J. Indian Math. Soc. (N.S.)* **20**, pp. 103–116.

Ribet, K. and Stein, W. (1999). Lectures on Serre's conjectures, in B. Conrad and K. Rubin (eds.), *Arithemetic Algebraic Geometry, IAS/Park City Mathematics Institute Lecture Series*, Vol. 7, pp. 143–232.

Ribet, K. A. (1977). Galois representations attached to eigenforms with Nebentypus, in *Modular functions of one variable, V (Proc. Second Internat. Conf., Univ. Bonn, Bonn, 1976)* (Springer, Berlin), pp. 17–51. Lecture Notes in Math., Vol. 601.

Rodriguez Villegas, F. (2007). *Experimental number theory, Oxford Graduate Texts in Mathematics*, Vol. 13 (Oxford University Press, Oxford), ISBN 978-0-19-922730-3.

Rose, H. E. (1994). *A course in number theory*, 2nd edition, Oxford Science Publications (The Clarendon Press Oxford University Press, New York), ISBN 0-19-853479-5; 0-19-852376-9.

Sarnak, P. (1990). *Some applications of modular forms, Cambridge Tracts in Mathematics*, Vol. 99 (Cambridge University Press, Cambridge), ISBN 0-521-40245-6.

Selberg, A. (1956). Harmonic analysis and discontinuous groups in weakly symmetric Riemannian spaces with applications to Dirichlet series, *J. Indian Math. Soc. (N.S.)* **20**, pp. 47–87.

Serre, J.-P. (1973a). *A course in arithmetic* (Springer-Verlag, New York), translated from the French, Graduate Texts in Mathematics, No. 7.

Serre, J.-P. (1973b). Formes modulaires et fonctions zêta p-adiques, in *Modular functions of one variable, III (Proc. Internat. Summer School, Univ. Antwerp, 1972)* (Springer, Berlin), pp. 191–268. Lecture Notes in Math., Vol. 350.

Serre, J.-P. (1979). *Local Fields* (Springer, New York).

Serre, J.-P. (1985). Sur la lacunarité des puissances de η, *Glasgow Math. J.* **27**, pp. 203–221.

Serre, J.-P. (1987). Sur les représentations modulaires de degré 2 de $\mathrm{Gal}(\overline{\mathbf{Q}}/\mathbf{Q})$, *Duke Math. J.* **54**, 1, pp. 179–230.

Sesiano, J. (1982). *Books IV to VII of Diophantus' Arithmetica in the Arabic translation attributed to Qusṭā ibn Lūqā, Sources in the History of Mathematics and Physical Sciences*, Vol. 3 (Springer-Verlag, New York), ISBN 0-387-90690-8.

Shanks, D. (1993). *Solved and unsolved problems in number theory. 4th ed.* (New York, NY: Chelsea. xiii, 305 p. $ 28.00).

Shimura, G. (1966). A reciprocity law in non-solvable extensions, *J. Reine Angew. Math.* **221**, pp. 209–220.

Shimura, G. (1971). On elliptic curves with complex multiplication as factors of the Jacobians of modular function fields, *Nagoya Math. J.* **43**, pp. 199–208.

Shimura, G. (1973). On modular forms of half integral weight, *Ann. of Math. (2)* **97**, pp. 440–481.

Shimura, G. (1994). *Introduction to the arithmetic theory of automorphic functions, Publications of the Mathematical Society of Japan*, Vol. 11 (Princeton University Press, Princeton, NJ), ISBN 0-691-08092-5, reprint of the 1971 original, Kanô Memorial Lectures, 1.

Shimura, G. (1998). *Abelian varieties with complex multiplication and modular functions*, *Princeton Mathematical Series*, Vol. 46 (Princeton University Press, Princeton, NJ), ISBN 0-691-01656-9.

Shimura, G. (2002). The representation of integers as sums of squares, *Amer. J. Math.* **124**, 5, pp. 1059–1081.

Silverman, J. H. (1992). *The arithmetic of elliptic curves*, *Graduate Texts in Mathematics*, Vol. 106 (Springer-Verlag, New York), ISBN 0-387-96203-4, corrected reprint of the 1986 original.

Silverman, J. H. (1994). *Advanced topics in the arithmetic of elliptic curves*, *Graduate Texts in Mathematics*, Vol. 151 (Springer-Verlag, New York), ISBN 0-387-94328-5.

Stein, W. and Joyner, D. (July 2005). SAGE: System for algebra and geometry experimentation, *Communications in Computer Algebra (SIGSAM Bulletin)* Available from `http://sage.sourceforge.net/`.

Stein, W. A. (2007). *Modular Forms, a Computational Approach*, *Graduate Studies in Mathematics*, Vol. 79 (American Mathematical Society).

Stein, W. A. and Watkins, M. (2002). A database of elliptic curves—first report, in *Algorithmic Number Theory (Sydney, 2002)*, *Lecture Notes in Computer Science*, Vol. 2369 (Springer, Berlin), pp. 267–275.

Sturm, J. (1987). On the congruence of modular forms, in *Number theory (New York, 1984–1985)*, *Lecture Notes in Math.*, Vol. 1240 (Springer, Berlin), pp. 275–280.

Swade, D. (1995). The Phillips Economic Computer, *Resurrection — The Bulletin of the Computer Conservation Society* **12**, pp. 11–18.

Swinnerton-Dyer, H. P. F. (1973). On l-adic representations and congruences for coefficients of modular forms, in *Modular functions of one variable, III (Proc. Internat. Summer School, Univ. Antwerp, 1972)* (Springer, Berlin), pp. 1–55. Lecture Notes in Math., Vol. 350.

Taylor, R. and Wiles, A. (1995). Ring-theoretic properties of certain Hecke algebras, *Ann. of Math.* **141**, 2, pp. 553–572.

Uhler, H. S. (1956). Exact values of 996! and 1000!, with skeleton tables of antecedent constants, *Scripta Mathematica* **21**, pp. 261–268.

Verrill, H. (2001). Fundamental domain drawer, Available at `http://www.math.lsu.edu/~verrill/fundomain/`.

Wan, D. (1998). Dimension variation of classical and p-adic modular forms, *Inventiones Mathematicae* **133**, pp. 449–463.

Wells, D. (1998). *The Penguin Book of Curious and Interesting Numbers* (Penguin, London).

Whittaker, E. T. and Watson, G. N. (1996). *A course of modern analysis*, Cambridge Mathematical Library (Cambridge University Press, Cambridge), ISBN 0-521-58807-3.

Wikipedia (2007). `http://en.wikipedia.org/wiki/Fundamental_Domain`, Retrieved on 24th September 2007.

Wiles, A. (1995). Modular elliptic curves and Fermat's last theorem, *Ann. of Math.* **141**, 2, pp. 443–551.

Wilf, H. S. (2006). *generatingfunctionology*, 3rd edition (A K Peters Ltd., Welles-

ley, MA), ISBN 978-1-56881-279-3; 1-56881-279-5.
Wolfram, S. (1996). *The Mathematica book*, 3rd edition (Wolfram Media, Inc., Champaign, IL), ISBN 0-9650532-1-0; 0-521-58888-X.

List of Symbols

$\mathbf{P}^1(\mathbf{Z}/N\mathbf{Z})$	projective 1-space over $(\mathbf{Z}/N\mathbf{Z})$, page 22
$f[\gamma]_k$	variant notation for the weight k operator, page 29
$f \mid [\gamma]_k$	weight k operator, page 29
$f \mid_k [\gamma]$	variant notation for the weight k operator, page 29
$f^{[\gamma]_k}$	variant notation for the weight k operator, page 29
$M_k(\Gamma)$	space of modular forms for Γ of weight k, page 30
$S_k(\Gamma)$	space of cusp forms for Γ of weight k, page 30
χ	Dirichlet character modulo N, page 30
q_N	$e^{(2\pi i z)/N}$, page 30
$M_k(\Gamma, \chi)$	space of modular forms for Γ of weight k and character χ, page 31
$S_k(\Gamma, \chi)$	space of cusp forms for Γ of weight k and character χ, page 31
$\Gamma^\star(N)$	congruence subgroup of level N, page 31
$G_k^{(0,a_2)}(z)$	Eisenstein series of level $\Gamma_1(N)$, page 32
$G_k^{\mathbf{a}}(z)$	Eisenstein series of level $\Gamma(N)$, page 32
∂_k	map from $M_k(\Gamma)$ into $M_{k+2}(\Gamma)$, page 35
θ	differential operator on modular forms, page 35
$v_P(f)$	order of vanishing of f at the point P, page 42
$v_\infty(f)$	index of the first non-vanishing term in the Fourier expansion of f, page 42
$\left(\frac{\cdot}{p}\right)$	Legendre symbol modulo p, page 50
ϕ	Euler totient function, page 50
T_n	n^{th} Hecke operator, page 59
$\mathrm{GL}_2^+(\mathbf{Q})$	subgroup of elements of $\mathrm{GL}_2(\mathbf{Q})$ with positive determinant, page 63
$f \mid [\Gamma_1 \alpha \Gamma_2]_k$	double coset operator, page 63
$\langle d \rangle$	diamond operator, page 64
$\mathbf{T}_R(S_k(\Gamma))$	Hecke algebra acting on $S_k(\Gamma)$ over R, page 67
U_p	Hecke operator acting on $M_k(\Gamma_0(N), \chi)$ with p dividing N, page 68
V_p	map from $M_k(\Gamma_1(N))$ into $M_k(\Gamma_1(Np))$, page 69
$\langle , \rangle$	Petersson scalar product, page 69
$\overline{x}$	the complex conjugate of x, page 70
$S_k^{\text{new}}(\Gamma_1(N))$	space of newforms in $S_k(\Gamma_1(N))$, page 81
$S_k^{\text{old}}(\Gamma_1(N))$	space of oldforms in $S_k(\Gamma_1(N))$, page 81
f_χ	modular form f twisted by a Dirichlet character χ, page 83

$j(z)$ — modular j-invariant, page 95

$\eta(z)$ — Dedekind η function, page 98

$\Gamma_0(p)+$ — congruence subgroup containing $\Gamma_0(p)$, page 106

E — elliptic curve, page 116

$a_p(E)$ — integer given by $p+1- \mid E(\mathbf{F}_p) \mid$, page 118

A — square, symmetric, positive definite integer matrix of dimension d, page 122

Δ_A — differential operator acting on polynomials, page 122

$r_A(n)$ — number of solutions to $x^T Ax = n$, page 122

$\theta(z; h, A, N, P)$ — theta function, page 123

$r_m(n)$ — number of ways that we can write n as the sum of m squares, page 126

$M_{k/2}(\Gamma_0(4))$ — space of half-integral weight modular forms for $\Gamma_0(4)$, page 129

$M_{k/2}(\widetilde{\Gamma}_0(4))$ — variant notation for$M_{k/2}(\Gamma_0(4))$, page 129

$S_{k/2}(\Gamma_0(4))$ — space of half-integral weight cusp forms for $\Gamma_0(4)$, page 129

$S_{k/2}(\widetilde{\Gamma}_0(4))$ — variant notation for $S_{k/2}(\Gamma_0(4))$, page 129

$p(n)$ — number of partitions of n, page 135

$\mathcal{P}(n)$ — set of partitions of n, page 136

$M_k(\Gamma; R)$ — space of modular forms for Γ of weight k with Fourier coefficients in R, page 144

$S_k(\Gamma; R)$ — space of cusp forms for Γ of weight k with Fourier coefficients in R, page 144

$\mathbf{Z}_{(p)}$ — the local ring of $\mathbf{Z}$ at p, page 144

$\mathcal{M}_R$ — graded ring of all modular forms for $\mathrm{SL}_2(\mathbf{Z})$ with Fourier coefficients in R, page 144

$\mathcal{S}_R$ — graded ideal of all cusp forms for $\mathrm{SL}_2(\mathbf{Z})$ with Fourier coefficients in R, page 144

$M_k(R)$ — space of modular forms for $\mathrm{SL}_2(\mathbf{Z})$ which have Fourier coefficients in R, page 145

$\overline{M}_k$ — classical space of mod p modular forms of weight k, page 145

ρ_f — representation attached to the modular form f, page 153

$S_k(\Gamma; R)_{\mathrm{Katz}}$ — space of Katz cusp forms for Γ of weight k with Fourier coefficients in R, page 156

$O(x^n)$ — big-O notation, page 182

$\mathbf{Z}_p$ — the p-adic ring of integers, page 186

Index